1.1   What is occupational health?

The discipline of occupational health is concerned with the two-way relationship of work and health:

Health $\rightleftharpoons$ Work

It is as much related to the effects of the working environment on the health of the worker, as it is to the influence of the worker's state of health on his/her ability to perform the tasks for which s/he was employed. The main thrust of the discipline is to prevent ill health rather than to cure it.

A joint ILO/WHO Committee defined the subject in 1950 as 'the promotion and maintenance of the highest degree of physical, mental and social well-being of workers in all occupations'. The provision of such a service to the workforce requires managerial and union involvement. A large number of professionals are also involved, including

- physicians
- nurses
- occupational hygienists
- lawyers
- toxicologists

- health physicists
- microbiologists
- epidemiologists
- ergonomists
- safety engineers

Doctors and nurses undertake duties which overlap but are rarely capable alone of providing all that is necessary to correct a work-related disease. The sequence of events in the successful control of a recently discovered health hazard and the person(s) involved are plotted in Table 1.1.

Some or all of the professionals listed would be involved in control of a work-related illness. In addition, the epidemiologist may investigate populations of workers exposed to similar conditions in the index factory, as well as extending the study to similar plants. The toxicologist could

Table 1.1 Health hazard control

Event	Person(s) responsible for control
Recognition of health effect ↓	Worker/safety representative/nurse/doctor
Diagnosis of illness	Nurse/doctor
Treatment ↓ (possibly)	Doctor
Discovery of environmental cause ↓	Hygienist/nurse/doctor/toxicologist
Monitoring and control of cause ↓	Hygienist/safety engineer/ergonomist and/or doctor
Monitoring of health of workers	Nurse/doctor/epidemiologist/toxicologist

study the effects of a suspect chemical in tissue cultures or animal experiments. The plant management would be expected to take an active part in proceedings and, in particular, the personnel and production departments could well influence as be influenced by the sequence of events. Nevertheless, the main professions involved are the medical staff (doctors and nurses) and the hygienists. A brief review of the functions of the other groups appears in Section 1.5.

1.2   Occupational medicine

Occupational medicine is the clinical speciality concerned with the diagnosis, manangemênt and prevention of diseases due to, or exacerbated by, workplace factors. In 1978, the Royal College of Physicians of London established a Faculty of Occupational Medicine. This action not only gave the occupational physician official recognition of his/her status as a medical specialist but also provided the framework for regulating formal training in the speciality. Specialist accreditation in the United Kingdom is the responsibility of the Joint Committee for Higher Medical Training (JCHMT). Similar accreditation procedures exist in the North America (Board of Preventive Medicine certification) and most Western European countries. Occupational medicine is primarily a branch of preventive medicine, with some therapeutic functions. The main attributes and functions of a doctor working in industry are

- knowledge of the work environment
- clinical skill in the early detection of ill health
- knowledge of relevant legislation
- pre-placement, periodic and special medical examinations
- administrative responsibility for nurses and first-aiders
- treatment
- health education and health promotion
- rehabilitation
- teaching and research
- advice to individuals, management, organized labour and safety representatives
- maintenance and review of clinical and environmental records
- surveillance of groups at special risk, e.g. lead workers, compressed air workers, vocational drivers
- liaison with outside organizations—Government, universities, other industries.

POCKET CONSULTANT

Occupational Health

J.M. Harrington BSc MSc MD FRCP FFOM FACE
Professor of Occupational Health, Institute of Occupational Health
University of Birmingham

F.S. Gill BSc MSc CEng MIMinE FIOH Dip. Occ. Hyg. Hon. FFOM
Consultant in Occupational Hygiene

In collaboration with

Tar-Ching Aw MBBS MSc PhD FRCP(C) FFOM
Senior Lecturer in Occupational Medicine
Institute of Occupational Health, University of Birmingham

G. Applebey LLB MCL
Lecturer in Law, University of Birmingham

C.P. Atwell RGN OHNC Cert Ed
Head, Occupational Health & Safety Unit,
Birmingham City Council

Third Edition

b

Blackwell
Science

© 1983, 1987, 1992 by
Blackwell Science Ltd
Editorial Offices:
Osney Mead, Oxford OX2 0EL
25 John Street, London WC1N 2BL
23 Ainslie Place, Edinburgh EH3 6AJ
238 Main Street, Cambridge,
  Massachusetts 02142, USA
54 University Street, Carlton,
  Victoria 3053, Australia

Other Editorial Offices:

Arnette Blackwell SA
  1, rue de Lille
  75007 Paris
  France

Blackwell Wissenschafts-Verlag GmbH
  Kurfürstendamm 57
  10707 Berlin
  Germany

  Feldgasse 13
  A-1238 Wien
  Austria

First published 1983
Second edition 1987
Reprinted 1988, 1990
Third edition 1992
Reprinted 1993, 1994 (twice), 1995

Set by Semantic Graphics, Singapore
Printed and bound in Great Britain
at The Alden Press, Oxford

DISTRIBUTORS

Marston Book Services Ltd
PO Box 87
Oxford OX2 0DT
(*Orders*: Tel: 01865 791155
          Fax: 01865 791927
          Telex: 837515)

North America
Blackwell Science, Inc.
238 Main Street
Cambridge, MA 02142
(*Orders*: Tel: 800 215–1000
               617 876–7000
          Fax: 617 492-5263)

Australia
Blackwell Science Pty Ltd
54 University Street
Carlton, Victoria 3053
(*Orders*: Tel: 03 347-0300
          Fax: 03 349-3016)

British Library
Cataloguing in Publication Data

Harrington, J.M.
  Occupational health.—3rd ed.
  (Pocket consultant)
  I. Title   II. Gill, F.S.
  III. Series
  363.110941

  ISBN 0-632-03189-1

Contents

Preface to the third edition, iv

Acknowledgements, v

1   Introduction, 1

2   Health services, 13

3   Evaluating workplace hazards, 27

4   Occupational diseases, 81

5   Chemical agents, 123

6   Physical agents, 167

7   Biological agents, 207

8   Special issues in occupational health, 217

9   Control of airborne contaminants, 237

10   Personal protection of the worker, 271

11   Some legal background to occupational health, 295

12   Education, 319

13   Sources of information, 335

Index, 343

Preface to the third edition

The request from our publishers for a third edition was received with a mixture of pleasure and dread. We were pleased that our little book continues to provide a useful introduction to occupational health. There are many more small books on the market these days but we feel that our original aim, to combine occupational medicine and occupational hygiene in one compact volume, has proved to be successful and still appears to be unique.

In this edition we have taken the opportunity to revise thoroughly all sections of the book and to change the structure of some sections. Greater emphasis (and space) has been given to the evaluation of workplace hazards in an attempt to provide practical guidelines for the task. We have added new sections and revamped the control of workplace hazards and worker protection to include important new legislation. The law section itself has been widened to take a European perspective, and throughout we have tried to diminish the national insularity that so often mars otherwise good texts.

As a result, the book is longer but hopefully is still manageable as an introduction to the subject. However, reading about occupational health is no substitute for getting out into the workplaces, talking to those who earn their living there and then acting on the evidence acquired in order to prevent ill health among them.

Malcolm Harrington *Birmingham*
Frank Gill *Petersfield*

Acknowledgements

The dread we felt at the task facing us in revising and rewriting large sections of the book was also felt by our assistants. Nevertheless, George Applebey and Cynthia Atwell have, again, done an excellent job on the legal and nursing aspects of the book. In addition, we are indebted to Ching Aw for his contributions on the health care industry and repetitive strain disorders as well as for the time and trouble he took to read and criticize the whole text.

Caroline Baxter, Jayne Grainger and Jane Hill bore the brunt of our repeated changes of text and unreadable handwriting with stoicism and good humour. Theirs was a vital task but with little of the limelight. Our information officer, Christine Bashford, provided valuable assistance on sources of information. We are fortunate to have such admirable staff at the Institute.

Particular gratitude is owed to the Controller of Her Majesty's Stationery Office for permission to reproduce the material on noise and hood design.

1 Introduction

1.1   What is occupational health? 3

1.2   Occupational medicine, 4

1.3   Occupational hygiene, 5

1.4   Occupational health nursing, 6

1.5   Related disciplines, 8

1.6   Summary, 11

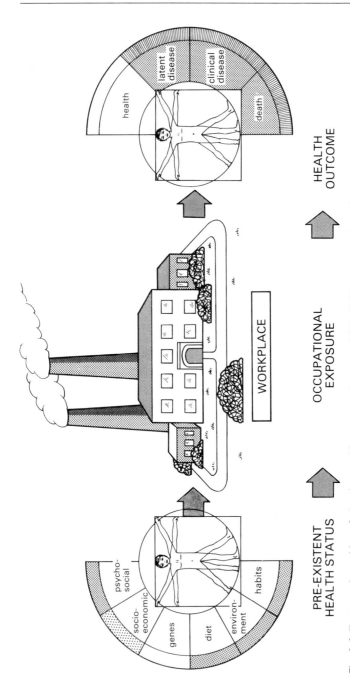

Fig. 1.1 Illustrates the problems facing the practitioner attempting to establish a link between work and health. The new employee brings a legacy of genetic, social, dietary and environmental factors affecting health to the new workplace which may influence his/her response to workplace hazards.

PRE-EXISTENT
HEALTH STATUS

OCCUPATIONAL
EXPOSURE

HEALTH
OUTCOME

WORKPLACE

psycho-
social

socio-
economic

genes

diet

environ-
ment

habits

health

latent
disease

clinical
disease

death

1.3 Occupational hygiene

This subject is defined in the handbook of the British Occupational Hygiene Society as follows: 'Occupational hygiene is the applied science concerned with identification, measurement, appraisal of risk and control to acceptable standards of physical, chemical and biological factors arising in or from the workplace which may affect the health or well-being of those at work or in the community.'

Table 1.2 lists these factors and also includes ergonomic and psycho-social ones. The latter tend not to be the province of the occupational hygienist; his main duties should include recognition and evaluation of a hazard and measures to control it.

Recognition of possible factors affecting health and comfort is achieved by studying work processes to ascertain the
- nature of the materials used
- products and by-products involved
- possible points of release or emission of hazardous agents
- posture and movements of the operatives
- hours and duration of rest periods at work
- nature of protective equipment provided.

Evaluation of the degree of hazard is gauged by
- measuring the intensity or concentration of the hazardous agent
- comparing the results against known standards or researched toxicological data
- ascertaining the human physiological effects upon workers from tests provided by medical sources, e.g. blood, urine analysis, lung function tests, nerve conduction velocities, etc.
- forming a judgement as to the degree of hazard and the possible

Table 1.2 Factors which affect workers' health

Physical	Chemical	Biological	Mechanical and ergonomic	Psycho-social*
Noise	Liquids	Insects	Posture	Worry
Vibration	Dusts	Mites	Movement	Work pressure
Ionizing radiation	Fumes	Moulds	Repetitive actions	Monotony
Non-ionizing radiation	Fibres	Yeasts	Illumination and visibility	Unsocial hours
Heat and cold	Mists	Fungi		
Electricity	Gases	Bacteria		
Extremes of pressure	Vapours	Viruses		

* See also Section 8.5.

remedies to the workplace environmental conditions.

The design of control measures to improve working conditions involves
• redesign of the work process and procedures
• substitution of safer materials
• attenuation of the intensity of the hazard
• shielding or screening of the worker against the hazard
• designing ventilation systems to extract or dilute airborne toxins
• conditioning the workplace
• drawing up work/rest regimes or job rotation to reduce worker exposure
• providing protective clothing.

Additional functions of the occupational hygienist should be to
• educate and train workers and management on the hazards of the work and on the use of the best procedures to minimize the risk
• prepare labels and precautionary texts for the safe handling of materials and physical agents
• assist in the design of new plant and modifications to existing plant to minimize risk to workers
• keep good records of all environmental measurements for future reference
• conduct research into specific health hazards that arise from the workplaces under their jurisdiction.

1.4   Occupational health nursing

The role of the occupational health nurse has changed rapidly over the past two decades. Originally, the nurse was employed in the organization to provide a treatment and first-aid service, dealing with accidents and illness at work. The profession of nursing has developed to encompass all aspects of preventive health care.

There is no legal requirement for organizations to employ occupational health nurses, although, with the introduction of new health and safety legislation which refers to the 'competent person' (e.g. COSHH Regulations), more organizations are employing nurses and expecting them to be 'competent' to carry out a full preventive role.

It is important for everyone to understand the role of the nurse in the workplace. It should encompass all those factors which may affect the health of people at work. The occupational health nurse must be pro-active and flexible in order to influence the health of employees and those within the wider sphere of the community outside the organization. The concepts of occupational health nursing should include the general environment, ecological, socio-economic and political factors which may affect occupational health practice, and should meet the needs of the employing organization, in order to improve and promote the health of its workers.

The qualified occupational health nurse may be an autonomous practitioner who can perform many functions either alone, or as part of a wider team of physicians, hygienists, safety officers, etc. S/He will carry out functions such as health supervision including pre-employment health assessment; follow-up assessments following illness or injury; assessment of those with known health problems and those working in potentially hazardous environments; development and implementation of immunization and vaccination programmes; hazard identification and control; counselling; health promotion and supervision and training of first-aid personnel.

The occupational health nurse is an 'adviser' to both management and employees; therefore it is imperative that the advice given is correct, is given in the right manner and is unbiased. There are approximately 9000 nurses working in the field of occupational health, the majority of which work without the support of a full-time medical officer; many have only a visiting GP with little or no occupational health training or experience. The United Kingdom Central Council (UKCC) for Nursing, Midwifery and Health Visiting and four National Boards were established by the Nurses, Midwives and Health Visitors Act 1979. The UKCC is the statutory regulatory body for the nursing, midwifery and health visiting professions. The Council and Boards provide a framework for developing nursing, midwifery and health visiting on a UK basis and for examining the needs and potentials of all areas of professional practice.

The National Boards, which were set up under the 1979 Act, are as follows:
The English National Board
The National Board for Scotland
The Welsh National Board
The National Board for Northern Ireland
(Please see Chapter 13 for full addresses.)

There are at present two levels of nurse: (i) RGN (Registered General Nurse), with three years' theoretical and practical training; and (ii) EN(G) Enrolled Nurse (General) with two years' mainly practical training. (EN training has now ceased—see Section 12.)

Any organization employing a nurse should ensure that the nurse's name appears on the professional 'register' or 'roll'. A nurse whose name appears on the 'register' or 'roll' is professionally accountable for his/her own

---

actions. The entry can be ratified by contacting the Registrar, UKCC for Nurses, Midwifery and Health Visiting, 23 Portland Place, London W1N 3 AF.

Failure to check the professional status of the nurse might have implications for management regarding the question of professional responsibility for the nurse's actions. In ensuring professional competence of the nurse, specialist education in occupational health will be necessary. (Refer to Section 12.3 for details.)

1.5   Related disciplines

Six important disciplines relating to occupational health are: law, toxicology, environmental engineering and epidemiology, which are covered at some length elsewhere in the text and ergonomics and safety engineering, which are expanded below.

In a place of work employing less than a few hundred employees, it is unlikely that the workforce will have access to more than a part-time doctor and/or nurse, though a hygienist might be used on a consultancy basis. The other disciplines could and, in some cases, should be involved in occupational health issues, though their functions may be subsumed in the main specialities.

Epidemiology

Epidemiology is the study of the distribution and determinants of disease frequency in human populations. As such, it may be readily applied to the field of occupational health, where various groups of people are employed in similar working environments. It is useful for the

- identification of new hazards
- control of known hazards
- establishment of hygiene standards to control or eliminate these hazards
- establishment of priorities in controlling a variety of hazards
- evaluation of health services designed to protect health and safety at work.

The starting point for many epidemiological studies is the collection of data relating to population health effects. These include

- death certificates
- birth certificates
- sickness—absence records
- industrial accident and injury claims
- general practitioner records
- hospital in-patient records
- pension scheme records
- professional association membership lists

1   Introduction

1.5   Related disciplines

- *ad hoc* collections of morbidity data
- establishment of computerized health records linked to environmental data.

The working environment itself is usually less well documented, particularly, in the past, in the case of factory working conditions. Nevertheless, in the best of all worlds, the following information should be available for a potentially toxic material

- personal and area samples
- variations in concentrations with work cycle, day, week, and specific operation
- high analytical accuracy
- low intra- and inter-laboratory and instrument variation. A brief description of epidemiological methods is outlined in Section 3.6.

Ergonomics

Ergonomics is the study and design of the working situation in order to benefit the worker-user. It is an attempt to match the machine to the worker rather than assume that the worker should match the machine and the environment. Assessment of the suitability of the relationship between work and worker requires examination of a number of factors, listed in Table 1.3.

The objective of ergonomics is to provide a satisfactory environment in which the worker can undertake the task required without undue physical or mental strain. Too frequently, workers are expected to operate machines with poor controls, sited at inappropriate heights or distances, in an uncomfortable environment. Such mismatching of person/machine can lead to disabling psychological or musculoskeletal disorders—see also Section 8.7. Furthermore, the ergonomist has an important role to play in *redesigning* machinery so that it can be operated by a disabled worker.

Table 1.3 Factors in assessing the working situation

Worker	Machine	Environment
Age	Size	Temperature
Sex	Purpose	Illumination
Race	Control devices:	Humidity
Body size and shape	knobs, levers, meters	Pressure
Energy expenditure	Frequency and complexity	Ventilation
Health status	of operation	Noise
Posture		Space
Movement		Relationship with other
Vision		workers and management

1.5   Related disciplines

Safety engineering

There should be a very close relationship between the safety engineering section of a company and the occupational health department. A safety policy actively pursued by vigilant staff can reduce the load on the medical services by reducing accidents, whilst a concerted drive by both safety and health staff helps to make employees aware of the dangers they face at work and to encourage them to look after their own health. Strong combined representations from both the medical and safety sides can bring pressure to bear on management to provide a workplace that is as safe and healthy as possible. In contrast to the long-term insidious aspects of occupational diseases, accidents at work require an immediate medical response. Factory surgeries need to be equipped for the treatment of injuries, which occur more frequently than industrial diseases. Where the risk of acute reaction to a chemical or physical hazard is high, due to accidental spills, leaks or other unintentional overexposure, the occupational health department needs to be aware of the risks and have emergency procedures and treatments ready. It is essential that the safety and health departments work closely together in this respect. The contribution from the union safety representative cannot be overestimated.

Unsatisfactory working environments can lead to a worker being under undue stress or below par, thus allowing his/her concentration or normal vigilance to lapse. This may lead to a greater risk of incurring an accident. Heat, cold or excessive noise may all contribute to such a lapse, as may ergonomic factors. Many solvent vapours also act as mild narcotics and have a similar effect on accident risk. A good occupational health department would normally advise safety engineers where such risks occur. The therapeutic (or illegal) use by the worker of psychotropic drugs can modify such responses to solvent exposure or may render the worker incapable of operating machinery in a safe manner.

Law

Like it or not, we are all governed by law, and this applies as much to our work as to other everyday activities. All forms of industrial activity have legal rules attached to them, and it follows that those professions involved with occupational health have to be aware of the legal framework within which they operate. *Ignorantia juris nemo excusat* ('ignorance of the law is no defence') is a maxim applying as much on the factory floor and in the consulting room as anywhere else. Certain areas of law must therefore be understood by those in the medical and related professions, often in precise detail, while other areas undoubtedly justify some basic knowledge, bearing

in mind that in cases of difficulty it is always advisable to seek proper legal advice. The main areas of law relevant to occupational health are
• health and safety legislation and its enforcement
• civil action for damages by persons suffering from occupationally related illness, disease or disability
• claim for disablement benefit under the social security system, usually for a 'prescribed disease'
• effect of illness or breach of safety rules on the contract of employment.
For a further account of law see Section 11.

1.6   Summary

Figure 1.2 provides an anatomy of occupational health procedures. It is not an area that is simply the province of the physician. Indeed, it is unlikely that the doctor alone could cope, despite the high regard frequently afforded him/her by industrial managers. S/he is part of a team of specialists concerned with the interrelationship of work and health. The objective is to ensure that the worker remains fit and well whilst undertaking the task s/he is employed to perform. To achieve this, the physician frequently relies on a nurse-based occupational health service aimed more at preventing ill health than curing it. S/he is usually dependent upon the hygienist for the measurement and control of environmental hazards in the workplace and may, in addition, require the services of safety engineers, ergonomists, toxiologists and epidemiologists to maintain the highest standards of physical and mental well-being in the workforce under his/her care. In all these endeavours the legal implications must be considered and a broad knowledge of the relevant law (Section 11) must be acquired.

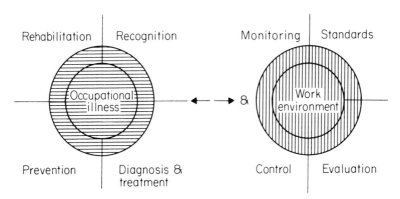

Fig. 1.2 Anatomy of occupational health procedures.

2 Health services

2.1 Introduction, 15

2.2 National health care in the UK, 15

2.3 Occupational health services in the UK, 17

2.4 International occupational health, 22

2.5 Summary, 25

2.1   Introduction

Occupational health services form only a small part of the health care facilities available in most Western countries. In Eastern Europe, however, occupational health is frequently an integral part of the state-controlled health services. In the past, assumptions were made that such extensive and well-financed services would lead to a high level of worker health and safety. The recent collapse of many totalitarian regimes has provided an opportunity to view the success or otherwise of occupational health. It seems clear now that the occupational and environmental health problems of Eastern Europe are vast—so much for state monopoly. In some Third World countries the services provided by a major employer of labour often encompass the workers' families and give 'cradle to the grave' cover. Otherwise, the state provides little cover for the workforce in the face of more acute problems of malnutrition and infectious disease control.

2.2   National health care in the UK

The National Health Service

Health care services in Britain are dominated by the National Health Service (NHS), which was introduced in 1948 in the face of much doubt and not without scorn. It was built on four fundamental principles:

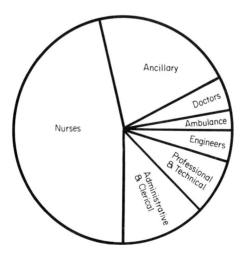

Fig. 2.1 Proportional distribution of NHS staff in the West Midlands in 1981. This regional health authority covers 10% of the UK population and employs 9% of the NHS staff.

- financing by taxes and contributions from the healthy for the benefit of the sick
- nationwide availability of high-quality service
- full clinical freedom for the doctors working in it
- focus on the family doctor team.

The NHS has matured over the years and, despite the criticisms and reorganizations (the latest is the two-tier regional/district system with the modifications and controls imposed by the new GP contracts plus the introduction of elements of privatization) it remains a model for many other countries. The present system employs a tiered structure of responsibility from Parliament through the Secretary of State for Health to the regional and district health authorities (RHA and DHA, respectively).

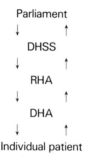

The NHS employs over 1 million people, ranging from ward cleaners to consultant medical practitioners. The majority of the employees of the service are nurses, who may work in a hospital, a family practice or the community (Fig. 2.1).

Environmental health

Environmental health is largely the responsibility of local government and is also organized on a regional and district basis. District health authorities have an obligation to provide expert staff and facilities, usually from among their community physicians, to local authorities to fulfil their environmental health responsibilities. The local authority, for its part, provides the formal environmental health departments with their environmental health officers—the community equivalent of the factory inspector. The recent Report on Public Health recommends a return to the old-style medical officers of health with a closer liaison between public health medicine and environmental health.

Private medicine
Public concern and disquiet in recent years about the efficiency of the NHS have greatly increased the growth and spread of private medical facilities, both for consultation and for hospital treatment. Private medical insurance schemes have burgeoned and the extension of industry-based group health insurance has made private medicine available to a wider range of people.

2.3   Occupational health services in the UK

The development of occupational health services in the UK has been largely through the *voluntary* provision of such services by employers. The early development of health care provision at work pre-dates any organized health care for the general public. It was not, however, the result of concern by the medical fraternity; early services were a manifestation of the philanthropic approach of certain enlightened factory owners such as Robert Owen and Quaker industrialists such as Cadbury and Fry. The services provided were general health care, with a provision for treatment services, neither of which were cheap or readily accessible to the nineteenth-century working classes.

Greater emphasis on prevention began in the first half of the twentieth century and measures to protect health and control the working environment were then grafted on to existing treatment-oriented services. The establishment of the Welfare State after 1945 led to a reappraisal of the need for nationalized occupational health services. Indeed, the Gowers Committee (1948), Dale Committee (1949–57), Alexander Working Party (1961) and Porritt Committee (1962) all urged such an extension of existing services. The Robens Committee (1970–72) took a different view, though the TUC's booklet on the subject (1981) still advocates nationalization.

Today, occupational health coverage is still patchy. A survey in 1976 commissioned by the Employment Medical Advisory Service highlighted this inadequacy. Indeed it is often still not clear how many doctors are involved in occupational health sessions.

The latest survey (1984) of occupational health and hygiene services in the UK was undertaken by the House of Lords Select Committee on Science and Technology. They confirmed the patchy nature of health coverage particularly for the small to medium sized workplace. The Committee's key recommendation for extending services was that the Health and Safety Commission (HSC) should instruct the Health and Safety Executive (HSE) to draw up a voluntary code of practice setting out the

2   Health services

2.3   Occupational health services in the UK

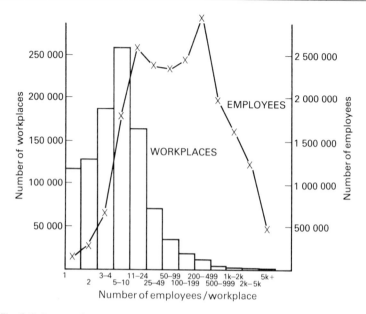

Fig. 2.2 Census of workplaces and populations employed (UK, 1981).

kind of service that should be provided in various types of industry. Despite government indifference, the HSC has embarked on such an endeavour. Figure 2.2 shows the large number of workplaces that employ small numbers of people. An expanding area of occupational health services in the UK is in the National Health Service—the largest employer of labour.

Aims and functions of an occupational health service
The aims of an occupational health service are laid down in the International Labour Organization's Recommendation No. 112 (1959), which was endorsed by the European Economic Community (1962) and the Council of Europe (1972). These aims have been endorsed in ILO Convention 161 and Recommendation 171 (1985). They are
• protection of workers against health hazards at work
• adaptation of the job to suit the workers' health status
• contribution to the establishment and maintenance of the highest degree of physical and mental well-being in the workforce.
    The basic functions of such a service are laid down in Part II Article 5 of Convention 161 (1985):

18

2.3 Occupational health services in the UK

Without prejudice to the responsibility of each employer for the health and safety of the workers in his employment, and with due regard to the necessity for the workers to participate in matters of occupational health and safety, occupational health services shall have such of the following functions as are adequate and appropriate to the occupational risks of the undertaking
- identification and assessment of the risks from health hazards in the workplace
- surveillance of the factors in the working environment and working practices which may affect workers' health, including sanitary installations, canteens and housing where these facilities are provided by the employer
- advice on planning and organization of work, including the design of workplaces, on the choice, maintenance and condition of machinery and other equipment and on substances used in work
- participation in the development of programmes for the improvement of working practices as well as testing and evaluation of health aspects of new equipment
- advice on occupational health, safety and hygiene and on ergonomics and individual and collective protective equipment
- surveillance of workers' health in relation to work
- promoting the adaptation of work to the worker
- contribution to measures of vocational rehabilitation
- collaboration in providing information, training and education in the fields of occupational health and hygiene and ergonomics
- organizing of first aid and emergency treatment
- participation in analysis of occupational accidents and occupational diseases.

The European Community Framework Directive (1989) (Section 11) goes, if anything , further than the ILO statements. For example: 'The employer shall designate a responsible person from supervisory staff or enlist services of competent outside agencies or individuals. The numbers and level of training will be determined by Member States taking into account the size of the undertaking and the hazards to which the employees are exposed.'

Similar implications for 'competent' health and safety surveillance are contained in the British Control of Substances Hazardous to Health Regulations 1988 (Section 3.2).

These recent changes in British and European health and safety rules are swingeing. The full impact of what they will mean has yet to be felt, or, in the case of some employers, yet to be realized!

Types of service

Full-time employees
A large petrochemical company, such as Esso, or a nationalized industry, such as British Coal, have a complement of full-time occupational physicians, nurses and hygienists, based centrally and locally. They may, in

---

addition, employ part-time practitioners. There is an integrated health and safety policy with regular liaison between health service staff and safety professionals. The Director of Services reports directly to the main board of the company and a senior manager ensures that agreed health and safety policy is implemented in the operating companies.

Part-time employees
Many companies employing 500 persons or less have a more rudimentary service, with perhaps a full-time nurse, a part-time doctor and a designated safety officer. Here, the service frequently reports to too low a level of management. Environmental monitoring may be undertaken by the safety officer or, occasionally, an outside consultant. A written health and safety policy *must* exist but senior management may play a lesser role in monitoring the implementation of such policy documents.

Group health services
Eleven such group services exist—the first was established at Slough in 1946 with Nuffield Foundation funds. The largest service covers 40 000 employees and, in all, 120 000 employees are covered by such services. In these schemes, the employer pays an affiliation fee in return for services, which may include an advisory service on health, safety and environmental problems, regular visits by nurses and less frequent visits from doctors, first-aid training, pre-employment and periodic health screening, supply of dressings and medications.

The size of various occupational health undertakings thus varies greatly. The Post Office employs a quarter of a million people and the Civil Service over half a million. The National Health Service is twice as large again, yet it has a less effective service due to regional and district variations in commitment and financing. Even so, the NHS employs 60 consultants, 300 full-time nurses and 1200 part-time doctors.

The cost per employee per year similarly varies. The group occupational health services try to provide a service for as little as £20–30, whereas some of the more sophisticated services to large private industries cost in excess of £100 per person per year. It is clear, however, that even the most rudimentary serivces are not financially viable below £10, although, at the other end of the scale, the sky is the limit.

Government agencies
See Employment Medical Advisory Service and Health and Safety Executive (section 11).

Occupational health ethics

The company doctor has two, sometimes conflicting, duties: one to the patient and the other to his employer. Nevertheless, certain criteria have been established to guide the physician in this potentially difficult position.

1   The occupational physician is an independent professional adviser to the employing organization.

2   Medical records are confidential to the health staff and the individual patient.

3   Medical records remain securely located in the health centre.

4   Certificates of fitness or otherwise, issued for management, contain no medical details unless written informed consent of the worker has been obtained.

5   Biological monitoring tests are available and explained to the individual concerned, whereas grouped, anonymous, test results are available to management and unions alike.

6   The doctor's responsibility to a worker exposed to a hazard takes precedence over the management's concern for commercial secrecy.

7   Research work may only be undertaken with the informed consent of the *individual* workers. Approval on their behalf by management or trade unions is not enough.

The future of occupational health services in the UK

The current haphazard growth of occupational health services is largely due to the varied reasons for their existence, which include

• philanthropic employers
• reduction of absenteeism due to sickness
• limitation of lost time and compensation claims
• an added company 'bonus'
• Government pressure
• union pressure
• hazardous processes

What is now needed is a more rational approach, given that scarce financial and staffing resources render impractical any plans for a nationalized service. For example, to provide the sort of service available to large-scale enterprise on a universal basis, would entail the employment of 10 000 doctors and 50 000 nurses. Some or all of the following improvements would provide a better organized and more broadly distributed service than currently exists:

• worker/management collaboration for established company health priorities

• regular industrial hygiene and ergonomic review
• effective first aid and minor injuries service
• basic health screening of new employees with periodic review
• industry-specific centres for information, advice and research
• professional bodies recognized as responsible for professional training
• regionally based specialist units with medical, hygiene and nursing
expertise—these could be based at a university or with the health authority
• central and regional reporting of occupationally related accidents, injury
and illness.

2.4   International occupational health

The latest ILO Convention (1985) states in Article 3(i) that each member
should undertake 'to develop progessively occupational health services for
all workers...in all branches of economic activity and all undertakings. The
provision made should be adequate and appropriate to the specific needs
of the undertaking.'

Europe

In terms of occupational health, Europe is in a state of flux. In the EEC, the
advent of the Framework Directive in 1989 lays the foundation upon which
various 'daughter' directives will be promulgated. The next few years will
be witness to a considerable number of specific measures aimed at
specific hazards or processes. At the same time, a new European list of
compensatable occupational diseases is due to be adopted by 1992 and
moves are afoot to establish basic guidelines for occupational health
training requirements. Standard setting (Section 3.5) will similarly become
more uniform. The national variation, therefore, will diminish as the will and
desire for 'harmonization' grow.

Individual national variations will still exist as the means to the end will
be left to individual member states. It is possible, therefore, that the annual
medical examination for all French employees will remain but other
methods of medical surveillance will be preferred in other countries. For
example, recent Irish legislation on health and safety is more extensive and
explicit than the British Health and Safety at Work Act; Dutch
compensation for occupational diseases is subsumed under their general
provisions for ill health and accident pensions whilst in the newer members
of the Iberian peninsula much remains to be done in the terms of basic
health and safety law.

2.4   International occupational health

In Eastern Europe, the reunification of Germany and the emerging democracies in other countries of the now defunct Warsaw Pact may lead to an extension of the EC to them as well as to Scandinavia. Whilst the Nordic countries enjoy the greatest degree of occupational health and safety provision of any European region, the problems of Eastern Europe are immense. Widespread and uncontrolled industrial pollution compounds the problems of outmoded factory machinery and overmanning. The scale of the task to bring all up to even the current EC average is daunting.

Worldwide

USA

There is a variable quality and quantity of occupational health services at factory level. The passage of the Occupational Safety and Health Act (1970) marked the watershed in American occupational medicine. Regulatory and research agencies were developed from the Act but, surprisingly perhaps, through separate Government departments. The Occupational Safety and Health Administration (OSHA) is responsible to the Department of Labor, whilst the research and service wing—the National Institute for Occupational Safety and Health (NIOSH)—is responsible to the Department of Health through the Centers for Disease Control of the Public Health Service. Variable degrees of co-operation and liaison exist between OSHA and NIOSH and with the Environment Protection Agency. The heads of all these agencies are normally appointed by the presidential administration, though an increasingly vociferous Congress plays an ever larger role in occupational and environmental health. OSHA inspectors and NIOSH investigators have similar powers of entry to their British counterparts, though they exercise these powers more frequently. Nevertheless, industry more frequently resorts to the courts to defend itself. There is much university-based research funded by NIOSH or through the relevant industry itself.

Canada

Here, developments are along similar lines to the USA but some years behind. Federal and provincial governments are currently spending considerable sums of money in establishing occuaptional health services and research programmes.

Japan
There are similar developments to those in Western Europe. The Industrial Health and Safety Law passed in 1972 (with amendments in 1975 and 1977) provides the power for Government to make regulations regarding safety and health surveillance for particular industries. Medical examinations are, in some instances, compulsory and doctors undertaking such work must be appropriately qualified.

Australia
Australia has developed a National Institute for Occupational Health and Safety as part of the Work Safe Australia programme which followed the 1985 legislation. The initial plans have been trimmed somewhat in the face of economic stringency but, in principle, the Federal Agency in concert with growing state programmes should provide the continent with an occupational health and safety structure similar in size and importance to the EC scheme.

The Third World
Occupational health is an increasingly important issue as rapid industrialization and mechanization proceed. Nevertheless, the predominant industry is still agriculture and the predominant health problems are still malnutrition and infectious disease. The problems of translocated migrant workers add to the social, medical and economic difficulties of the rapid growth of industrial cities. Legislation is sparse and rarely enforceable to the same degree as in the developed world.

High-class occupational health services, extended to include family health, are available to employees of the large multinationals operating in these countries. The World Health Organization has been particularly active in supporting the development of occupational health services and in the training of occupational health personnel. Well-equipped and competent centres for training, advice and research now exist in India, Sri Lanka, Singapore, Egypt and Nigeria. The needs for the future include
• more field-work to identify and quantify the size of the work-related health issues
• more training of personnel, preferably at local centres
• the provision of effective and enforceable occupational health law
• the incorporation of occupational health into an expanded rural and suburban health service programme.

2.5   Summary

Occupational health services are not universally available, nor are they interpreted in the same way within countries or between countries. Many such services began through philanthropic self-interest on the part of employers, whilst organized labour, national and state governments and the medical profession itself have been slow to extend such services. In most Western countries, occupational health has developed separately from mainstream medicine, although this was not the case in Eastern Europe. Many industrialized countries now have effective health and safety legislation which is, in most cases, little more than a decade old. Increased harmonization is to be expected on a Pan-European scale. Developing countries lag behind and have additional problems because, even before the worker is exposed to an industrialized working environment, s/he may be in poor health due to malnutrition, social migration and endemic, infectious or parasitic disease.

3 Evaluating workplace hazards

3.1   Walk-through surveys, 29

3.2   The Control of Substances Hazardous to Health Regulations
      1988, 32

3.3   Periodic medical examinations, 45

3.4   Occupational hygiene survey techniques, 57

3.5   Occupational hygiene standards for airborne contaminants, 59

3.6   Health survey design—the epidemiological approach, 62

      Appendix 3.1 Pre-employment medical forms, 66

      Appendix 3.2 Occupational history pro-forma, 73

      Appendix 3.3 Medical questionnaires, 74

      Appendix 3.4 Routine medical examinations, 78

3   Evaluating workplace hazards

The keywords to the discipline of occupational health are:
* recognition
* evaluation
* control.

   This chapter should start the reader on the road to the first two of these. The art of recognition is used in the walk-through survey where the practitioner should be alert to the potential of the working environment to cause ill health. Once the possible causes are recognized, they need to be evaluated so that they can be either eliminated as being a perceived hazard not borne out by subsequent investigation or accepted as a real hazard for which some degree of control will need to be implemented.

   Evaluation is an integral part of the Control of Substances Hazardous to Health Regulations 1988 as there is the requirement to assess the working environment. Assessment techniques are discussed in Section 3.4. Evaluating the obvious hazard may be by inspection but where airborne pollutants are concerned evaluation may involve occupational hygiene surveys, the results of which need to be judged by reference to the published standards. A brief discussion on surveys and sources of published standards are given here but other sources of information and techniques of control are covered elsewhere in this book. When continuing health hazards are present but controlled, the need to keep a watchful eye on the workforce is important; the role of periodic medical examination for that purpose is also discussed.

3.1   Walk-through surveys

One of the initial requirements of anyone evaluating workplace hazards is to go and see for oneself by undertaking a walk-through survey. In order to get the best out of such a visit, observation should be undertaken in an objective way, using a check list for guidance. This is probably best handled using a pro-forma, filling in the details as the survey progresses.

   A suggested pro-forma is given below.

Name of company

Address of site

Location of workplace

Name and designation of person responsible for the workplace

---

3   Evaluating workplace hazards

3.1   Walk-through surveys

Sketch plan of area surveyed (show positions of sources of exposure, locations of workers, ventilation extracts, etc.)

Item to be checked	Description or comment	Satisfactory (yes/no)	Action required
Numbers of people exposed (distinguish the numbers of each sex)			
Duration of shift and shift pattern			
Skill levels required and degree of training			
Degree of supervision			
CHEMICAL/BIOLOGICAL AGENTS Hazardous substances used, give names of substances and their form (dust, fibre, liquid, gas, vapour, microorganism) (if large list, append)			
Raw materials			
Final products			
Intermediate products			
Hazard data sheet available? (if yes, append)			
Route(s) of entry (inhalation, ingestion, skin contact, inoculation)			
Degree of exposure (subjective opinion or give results of monitoring)			

3   Evaluating workplace hazards

3.1   Walk-through surveys

Item to be checked	Description or comment	Satisfactory (yes/no)	Action required
Means of control (e.g. local exhaust ventilation, protective clothing, enclosures, screens, etc.)			
Method of monitoring performance and maintenance of control measures			
PHYSICAL AGENTS Hazardous agents present (e.g. noise, radiation, heat)			
Methods of control (shielding, enclosures, protective clothing)			
Method of monitoring performance and maintenance of control measures			
LIGHTING Give subjective impression or results of recent measurement			
GENERAL Written work procedures, do they exist?			
Housekeeping and management attitudes to health and safety (give subjective impression)			
HEALTH AND WELFARE medical, nursing, first aid facilities			
Wash rooms/showers rest rooms			
Clothing issue and laundry facilities			
Eating/drinking facilities			
Smoking policies			
Company health promotion policy			
Pre-employment and periodic health examinations			
Rehabilitation and disabled persons' policy			
Name of surveyor			

This should not be regarded as a comprehensive list, the surveyor is advised to leave space at the end of the forms for adding extra questions which will, no doubt, arise during the survey.

Having completed the initial assessment it should be possible to draw up a plan of action to be discussed and agreed with the managers of areas under scrutiny. Time limits should be set on any action and dates for follow-up surveys agreed.

3.2   The Control of Substances Hazardous to Health Regulations 1988

They came into force on 1 October 1989.

The following is a summary of the essential sections of the COSHH Regulations 1988. Note: *The wording below is by the authors and not that of the regulations as published.*

Abbreviations used in this text:

COSHH—control of substances hazardous to health

MEL—maximum exposure limit

OES—occupational exposure standard

EMA—Employment Medical Adviser

CP&L—classification, packaging and labelling

HSE—Health and Safety Executive

AFARP—as far as is reasonably practicable

ARP—as reasonably practicable.

Regulation 2. Interpretation

'Substance hazardous to health' means any substance that creates a hazard to health of any person arising out of or in connection with work which is under the control of the employer and includes:

• any substance listed in part 1A of the approved list of the CP&L Regulations 1984 and for which the classification is specified as: very toxic, toxic, harmful, corrosive, irritant

• any substance that has an MEL or OES

• a microorganism which creates a hazard to health

• dust of any kind when present in substantial concentrations in air.

Regulation 5. Application of Regulations 6–12

Exceptions are where other Regulations apply, namely: Control of Asbestos at Work Regulations 1987, Mines (respirable dust) Regulations 1980, Control of Lead at Work Regulations 1980; hazardous substances due to radiation, explosive or flammable properties or if it is solely at high or low pressure; where the substance is administered in the course of medical treatment.

Regulation 6. Assessment of health risks created by work involving substances hazardous to health
An employer shall make a suitable and sufficient assessment of the risks to the health of workers exposed to substances hazardous to health with a view of controlling those hazards.

Regulation 7. Control of exposure to substances hazardous to health
Every employer shall ensure that exposure of employees is prevented or adequately controlled. Where an MEL exists exposure should be reduced AFARP below that limit. Where an OES exists exposure should be: (i) not exceeded; or (ii) if exceeded, the reasons identified and appropriate action taken to remedy the situation as soon ARP. Control should not be achieved by resorting to respiratory protective equipment but if it has to be used it shall be suitable for the purpose and HSE-approved.

Regulation 8. Use of control measures
Every employer shall ensure that the control measures or protective equipment is properly used and every employee shall make full and proper use of the control measures and equipment provided and report any defects.

Regulation 9. Maintenance of control measures, etc.
Every employer shall ensure that the control measures provided be maintained in an efficient state and working order and in good repair.
    Where local exhaust ventilation is provided its performance shall be examined and tested at least every 14 months, except for certain specified processes: as indicated in Schedule 3 of the Regulations, and for any other process at suitable intervals. Where respiratory protective equipment is provided it shall be examined and tested at suitable intervals. Records shall be kept for five years of any test made under this regulation.

Regulation 10. Monitoring exposure at the workplace
Wherever necessary to maintain adequate control of exposure, to protect the health of employees or for substances listed in Schedule 4, regular monitoring is required, the records of which shall be kept for at least five years or, if required to be kept with medical records as described in section 11, then records shall be kept for thirty years.

Regulation 11. Health surveillance
• Where appropriate for their protection, employees exposed to substances hazardous to health shall have regular and suitable health surveillance if exposed to a substance or if engaged in a process listed in Schedule 5, or if an identifiable disease or ill effect may occur, or if there are valid techniques for the early detection of disease or ill effect.
• The employer is to keep an approved health record of those employees for thirty years.
• Where an employee is exposed to a substance listed in Schedule 5, part II, health surveillance shall include medical surveillance by an EMA or appointed doctor.
• The frequency of medical surveillance is specified in Schedule 5, part II, and can continue for a specified period after exposure has ceased.
• An EMA or appointed doctor may inspect any workplace in order to carry out his functions under this regulation.
• An EMA or appointed doctor can prevent an employee from working with that substance or s/he can lay down conditions under which that employee shall work.
• An employee shall present himself for medical examinations which shall be during working hours and at the employer's expense and if on the employer's premises shall be in suitable accomodation.
• An employee may see his/her own medical record and if aggrieved by what it contains may apply for a review.
• Medical records shall be made available to an EMA or appointed doctor as s/he may reasonably require.

Regulation 12. Information, instruction and training for employees who may be exposed to substances hazardous to health
• An employer who undertakes work which may expose an employee to SHH shall provide such information, instruction and training as is adequate for him or her to know: the nature of the substance and the risks to health created, the precautions which should be taken, the results of any monitoring and if the MEL has been exceeded and the collective results of any health surveillance undertaken under section 11.
• Every employer shall ensure that any person who carries out work in connection with duties under these Regulations is given such information, instruction and training as will enable him or her to carry out that work effectively.
    The Approved Code of Practice explains in some detail how the COSHH regulations should be complied with.

3   Evaluating workplace hazards

3.2   COSHH Regulations 1988

Summary of schedules referred to above

Schedule 3   Frequency of thorough examination and test of local exhaust
ventilation plant used for certain processes

Process	Minimum frequency
Processes in which there is blasting in or incidental to metal castings, in connection with their manufacture	Monthly
Processes, other than wet processes, in which metal articles (other than of gold, platinum or iridium) are ground, abraded or polished, using mechanical power in any room for more than 12 hours in any week	6-Monthly
Processes giving off dust or fume in which non-ferrous metal castings are produced	6-Monthly
Jute cloth manufacture	Monthly

Schedule 4   Specific substances and processes for which monitoring is
required

Substance or process	Minimum frequency
Vinyl chloride monomer	Continuous or in accordance with a procedure approved by the Health and Safety Commission
Vapour or spray given off from vessels at which an electrolytic chromium process is carried on, except trivalent chromium	Every 14 days

3.2   COSHH Regulations 1988

Schedule 5   Medical surveillance

Substances for which medical surveillance is appropriate	Processes
Vinyl chloride monomer	In manufacture, production, reclamation, storage, discharge, transport, use or polymerization
Nitro or amino derivatives of phenol and of benzene or its homologues	In the manufacture of nitro or amino derivatives of phenol and of benzene or its homologues and the making of explosives with the use of any of these substances
Potassium or sodium chlorate or dichromate	In manufacture
1-Naphthylamine and its salts *Ortho*-tolidine and its salts Dianisidine and its salts Dichlorbenzidine and its salts	In manufacture, formation or use of these substances
Auramine Magenta	In manufacture
Carbon disulphide Disulphur dichloride Benzene, including benzol Carbon tetrachloride Trichlorethylene	Processes in which these substances are used, or given off as vapour, in the manufacture of indiarubber or of articles or goods made wholly or partially of indiarubber
Pitch	In manufacture of blocks of fuel consisting of coal, coal dust, coke or slurry with pitch as a binding substance

COSHH assessments

It could be argued that the chemical/biological agents section again of the walk-through survey described in Section 3.1 could be replaced by the assessment required under these regulations. A pro-forma approach is again necessary for these assessments and the one developed by our Institute is given below.

3 Evaluating workplace hazards

3.2 COSHH Regulations 1988

ASSESSMENT FORM

Company name..

Works address ....

Location of workplace ....

Single workstation/process assessment   One substance/more that one

Number of workers exposed...

List of substances used:

Name (trade and IUPAC) ....

Physical state(dust, fibre, gas, etc.)..

Mode of exposure (inhalation, skin) ....

Toxicity class (very toxic, etc.) ....

Standard ....

Occupational exposure ....

Toxic effects of each substance (brief description of chronic and acute effects on target organs) ....

Sketch and/or flow chart of process if relevant (use separate sheet if necessary)

SOURCES OF EXPOSURE

(in the descriptions below outline how the substances(s) come into contact with the workers)

Storage (description to include type of containers, location, method of opening; also how stores issues are controlled) ....
....

Are leaks possible?   Yes/no   If yes give method of prevention if any................
....

Packaging and labelling

Is suitable packaging and labelling provided?   Yes/no
If no state what improvements should be made..
....

3.2   COSHH Regulations 1988

Transport and transfer (describe how substances are moved from store to point of use) ....
....

Is inhalation or skin contact possible?   Yes/no   If yes state which and describe the method of control if any ....
....

Are spills possible?   Yes/no   If yes state how and method of control if any ......
....

Use (describe how the substance is used; refer to the sketch where necessary).........
....

Is inhalation or skin contact possible?   Yes/no   If yes state how and give method of control if any ....
....

Disposal of excess material (describe how disposal is achieved).............................
....

Is inhalation or skin contact possible?   Yes/no   If yes state how and give method of control if any ....
....

Emissions of atmosphere (describe what is likely to be present in any emission to outside atmosphere from within the building)...
....

Are these emissions likely to cause any environmental problems?   Yes/no
If yes state how they can be minimized..
....

Waste products (describe what products and how they are disposed of; include these products in the list of substances above)...
....

Is inhalation or skin contact possible?   Yes/no   If yes state how and what method of control is used if any ....
....

Intermediate products (list any intermediate products that might occur and state where they could be inadvertently emitted into the workroom; also include these products in the list of substances above)...
....

Is inhalation or skin contact possible from these fugitive emissions?   Yes/no
If yes state how their effect could be minimized...
....

---

3 Evaluating workplace hazards

3.2 COSHH Regulations 1988

MONITORING

Workplace monitoring

Are airborne concentations monitored?   Yes/no   If yes state frequency ....
....

If no state whether measurements should be taken; give details ....
....

Give results with dates; if frequently or routinely give the results of the last three
surveys and state reference number of the appropriate reports or results sheets
(append extra sheets if necessary)..
....

Do the results above show that a hazard to health exists?   Yes/no   If yes give
details ....
....

Are surface contamination measurements necessary?   Yes/no   If yes give
details ....
....

Health/medical surveillance

Is health/medical surveillance undertaken?   Yes/no   If yes give or append
collective results ....
....

If no state whether surveillance should be undertaken and give details ....
....

Biological monitoring

Are biological measurements taken?   Yes/no   If yes state what and give
reference numbers of record and summary of results; do not mention individuals.......
....

Do results of health/medical surveillance or biological monitoring show any risk to
health?   Yes/no   If yes give details ....
....

CONTROL

Ventilation methods of control

If ventilation methods of control are used state frequency of routine measurement
and give reference number of record sheet ....
....

3   Evaluating workplace hazards

3.2   COSHH Regulations 1988

Do the results show any malfunctioning of the ventilation systems?   Yes/no
If yes give details ....
....

Protective equipment

If protective equipment is used described the type used and method of selection,
inspection and maintenance ....
....

Is protective equipment suitable and in good order?   Yes/no   If no give details
....
....

Is decontamination of protective equipment necessary?   Yes/no
If yes is it undertaken?   Yes/no   If yes give details ....
....

If no state what is required ....
....

Other methods of control not mentioned above

Give details of any method of control ....
....

Are these methods operating satisfactorily?   Yes/no   If no state what
improvements could be made ....
....

Training

Do any of the work methods described involve special training?   Yes/no
If yes give details ....
....

Is any training given with regard to the health and safety aspects of the work?
Yes/no   If yes give further details ....
....

Is this training adequate to minimize the health risk?   Yes/no   If no give details
of extra training required ....
....

Welfare and personal hygiene

List the provisions for welfare and hygiene ....
....

Are these provisions satisfactory?   Yes/no   If no state what improvements are
required ....
....

3   Evaluating workplace hazards

3.2   COSHH Regulations 1988

Health and safety work sheets

Are any health and safety work sheets issued?   Yes/no   If yes append a copy,
If no give details of what should appear on such a sheet; append a draft if possible

....

....

THE ASSESSMENT

Having considered the information provided on the previous pages, I am/we are of
the opinion that (tick as appropriate):

• risks to health are unlikely

• risk is significant but adequate controls are in operation

• risk is significant and controls need to be applied as follows: ....

....

• risk is unknown; the following actions are recommended: ....

....

This assessment should be reviewed (tick as appropriate):

• when the above actions are implemented

• when circumstances change

• ...... months from the date given below

Assessor(s):

Name .... Position .... Qualifications ....

Signature .... Date ....

Initial assessment procedures
In order to make a start on the assessment it is useful to have a sequence
of actions to follow.

1 *List the substances in the area to be assessed* This important first
stage helps to define the size of the task. If the list becomes very large
then the areas to be assessed should be reduced and subdivided into
manageable packages. A decision is required on whether a complete
production process is to be assessed or whether a sub-process within it is
more manageable. The number of individual substances appearing in any
one operation will probably be the deciding factor.

It is also important to determine the volume of storage and use of the substances under review.

2 *Determine which of those substances are actually used* This important consideration has proved to be a useful economic exercise in itself as many companies are finding their storerooms and cupboards well stocked with chemicals no longer used but not yet discarded. They are also finding that different sections of the plant are using different chemicals for the same or similar processes and that some rationalization of their purchasing policy is required which will financially benefit the company. The COSHH assessment has provided the ideal opportunity to remove all the old substances from the site, some of which may be in an unstable condition and others in containers that are deteriorating rapidly and may soon become an occupational or environmental danger.

3 *Determine their true chemical names and/or CAS numbers* Most substances appear in the workplace under a trade name or code number. If the toxic nature of the substance is to be determined from the standard texts then a precise identification is required. All chemicals are issued with a unique name by the International Union of Pure and Applied Chemistry (IUPAC) and a unique number known as the CAS or Chemical Abstracts Series number.

4 *Obtain suppliers' data sheets* There is a duty under section 6 of the Health and Safety at Work Act etc. 1974 for suppliers to provide adequate information on substances supplied and this is reinforced by the Consumer Protection Act 1987. This information is usually provided by the supplier in the form of a data sheet (see Hazard Data Sheets p. 44). The quality of the information supplied is very variable, the best giving all the information required to appraise the toxicity of the substance, the worst giving information that is misleading and sometimes dangerous.

It is advisable to have standard letters available to request this information and more strongly worded back-up letters in the event of default.

5 *Evaluate data sheets* It is wise to check the validity of the information supplied on the data sheets. For example, the IUPAC name of the substance or substances may not be given, making it difficult to check the toxicity information provided. Alternatively, if the substance is a mixture of chemicals such as a proprietary solvent, not all the substances may be shown. It is understandable if a supplier does not give the exact formulation of a mixture as s/he has 'trade secrets' to protect but a list of substances present without the exact proportions can be given without running the risk of industrial espionage!

6 *Check the toxicological data given and rewrite data sheet* Once the name of the substance is known a simple check on the accuracy of the toxicological data given should be made before writing the data sheet to suit the way the substance is used in the situation being assessed. The data sheet will need to be rewritten or supplemented to take into account the way the substance is to be used. The suppliers cannot be expected to anticipate the way their substance is to be stored, transported or handled in the workplace under review but the employees will require some guidance. This is a requirement of Regulation 12 of the COSHH Regulations (see Hazard Data Sheets p. 44).

7 *Inspect the places where the substances are handled* Now is the moment to inspect the way the substance is being handled to establish the modes of exposure and the possible risk to those employed. That is, the way the material is stored, transferred to the point of use, dispensed into the process and disposed of after use all poses a potential risk to those involved. It is in this way that it is possible to establish whether the exposure is to the skin or via inhalation, the two commonest modes. At the same time on-site observations can be made on the eating, drinking and smoking activities in the workplace, all of which can be a potential source of ingestion.

8 *Inhalation route—check airborne monitoring* If the substances are dusty or volatile and there are open containers providing surfaces for evaporation there is a likelihood of inhalation being the main route of entry. It will then be necessary to check the airborne concentration of the substances in the breathing zone of the worker and to compare the results with published standards (see standard setting, Section 3.5). Occupational hygiene surveys may need to be arranged.

9 *Skin contact route* Observations on the method of handling will reveal whether skin contact is likely. When liquids are being transferred from one receptacle to another, even if mechanically handled, splashing could occur and with any open surface of liquid accidental contact is possible. Also the handling of wet materials with unprotected hands is an obvious source of exposure. No measurements are adequate to establish the degree of exposure but the wary eye backed up with knowledge of the material's potential dangers may be what is needed to assess this hazard.

10   *Look at the method of control* The performance of control methods (Chapter 9) needs to be assessed. In some cases this can be done by observation, whilst for others it will involve some scientific measurements. If airborne substances are being controlled, then the ultimate test is the airborne concentration in the breathing zone of the workers involved. If the levels are substantially below an applicable MEL, or below the OES, then control will be achieved.

More subtle methods of control involving working procedures and good supervision will have to be checked and seen to be working satisfactorily before accepting that the process is free from risk.

11   *Implement improvements before the final assessment* If as a result of this initial assessment procedure some obvious faults are seen, then they should be speedily rectified before completion of the final assessment. If the improvements appear to require time to implement, then an interim assessment should be made with a view to reassessment later.

Hazard data sheets
The purpose of the data sheet is twofold:
• the receipt of product information from the supplier
• the provision of information to users within the company.
The latter is necessary in order to fulfil the company's obligations under Regulation 12 of COSHH. The sheets should not be the same for reasons given below.

A good suppliers' hazard data sheet should contain the following information:
• identification: product name, physical form, e.g. powder, liquid, etc., colour, odour
• supplier: name, address, emergency tel. no., contact person
• composition: chemical names of constituents, CAS nos., synonyms, proportions, formulae, occupational exposure limits, impurities
• physical data: boiling point, vapour pressure, specific gravity, melting point
• health hazards: short- and long-term effects of inhalation, skin contact, ingestion, injection, eye contact, first detectable signs of overexposure
• emergency and first-aid procedures

- spillage procedures
- fire precautions and likely products of combustion
- recommended control measures other than personal protective equipment
- recommended PPE
- storage, packaging and labelling advice
- reactivity data: stability, decomposition products, known interactions
- special precautions
- legal requirements
- sources of information.

Very few will be as comprehensive as this (or will need to be) if the substance poses a low risk. Users who are large consumers of a particular product may ask their suppliers to complete a company standard data sheet for circulation within the company and to be kept on an 'in-house' data base.

For 'shop floor' purposes, the suppliers' data sheet may contain too much information or be couched in unintelligible scientific terms. In addition, the sheet for shop floor use must contain details of safe systems of use and details of the local methods of control. For example, the solvent 1,1,1-trichlorethane is used in large quantities on the shop floor for degreasing and in small quantities in the office as a correcting fluid; the method of exposure and control will be different in each situation. A separate data sheet should be provided for the two types of user. Data sheets should also be written in language that the shop floor worker can understand and where necessary may have to be issued in several languages.

3.3   Periodic medical examinations

Introduction

Medical examinations figure prominently in the functions of occupational health personnel (Chapter 1) and, indeed, managers in industry are often keen that such services should be available to the workforce. Unfortunately, some occupational health services are little more than a glorified health screening device where little critical evaluation has been applied to the justification for such use of costly medical resources. Both management and labour tend to share the unsubstantiated belief that periodic medical examinations offer some preventive health advantage. Such preconceived notions may be supported by physicians who see useful sessional fees accruing from the performance of such manoeuvres which, in truth, vary in value from the marginally useful to the useless. The essential fallacy lies in the overestimation of the information obtainable from the routine medical examination, with or without support investigations such as chest radiography, electrocardiography and pulmonary function tests.

Moreover, often little attempt is made to differentiate *screening* from *case finding. Screening* may be defined as an activity making use of procedures by which unselected populations are classified into two groups: one with a higher probability of being affected by killing or disabling conditions and the other with a low probability. *Case finding* is the detection of disease by means of various tests or procedures by a health worker who has a continuing close relationship with the examinee. Recent reviews of the value of periodic medical examinations suggest that there are only a few adult conditions worth the trouble of systematic surveillance, either for prevention or treatment. These include

• communicable disease such as rubella, smallpox, cholera, tuberculosis, hepatitis, etc.
• visual defects such as refractive errors
• primary open-angle glaucoma
• diabetes mellitus
• hypothyroidism
• hypertension
• chronic bronchitis
• cancer of the breast
• cancer of the cervix
• cancer of the colon and rectum
• cancer of the bladder
• cancer of the skin
• dental caries.

The alteration of an individual's life-style or habits, e.g. controlling alcohol and cigarette consumption, is feasible and can be effective but has poor response rates in counselled individuals.

Periodic medical examinations do, however, form an important part of occupational health practice *if* they are undertaken for specific reasons, with specific objectives and with specific consequent action by the physician in the light of the results. Such examinations may be

• statutory—required by law, e.g. Lead Regulations
• voluntary—a requirement applied by the employer alone or on medical advice.

The common examinations are

• pre-placement
• periodic—post-sickness-absence, specific occupational groups, pre-retirement.

3 Evaluating workplace hazards

3.3 Periodic medical examinations

Pre-placement examinations
The reasons for these examinations may include
• assessment of fitness to do the job specified, e.g. heavy goods or public service vehicle drivers (statutory), food handlers (may be statutory), divers (statutory)
• assessment of fitness for *any* job
• detection of ill health which may be remedied to allow applicant to do specified job or job adjusted to allow performance by applicant in *current* state of health
• base-line information on fitness
• pension fund/insurance/superannuation criteria
• management demands
• review of disablement to enable suitable placement.

Much of the pre-placement assessment can be carried out by a competent nurse. Appendix 3.1 illustrates the procedure adopted in one large company. A health questionnaire is self-administered (Part 1), followed by a review by the nurse and some additional tests, such as blood pressure, urinalysis, vision, etc. (Part II). *In the absence of a specific need to see the physician,* this will suffice. If, however, the nurse feels that further medical opinion is necessary or the job or type of individual warrants it, the medical examination is undertaken (Part III) with or without additional investigations (Part IV).

In practice, the doctor rarely sees more than 10% of new applicants. Those s/he does see have been established, a priori, as necessary for review. These might include the following:
• employees concerned with public safety, e.g. drivers, airline pilots, food handlers
• employees where high non-specific demands are made on them in physically strenuous jobs
• employees where high specific demands are made on them.

Specific demands and the assessment of the hazards caused by them include
• dust—chest radiography
• noise—audiometry
• ionizing radiation—haematology
• organic solvents—serum biochemistry
• laboratory workers—immune status assessment
• allergens—atopy tests.

The doctor has an important role to play in educating managers in what are valid criteria for acceptance or rejection of an applicant on medical grounds. As a general rule, it should be possible to alter most jobs to suit the applicant who, after all, has been selected by the management for the post, pending the results of the medical. Nevertheless, muddled thinking on these issues is still prevalent, particularly where superannuation schemes are involved. For example, some employers will still not, on principle, employ an insulin-dependent diabetic, no matter how well controlled. The convoluted reasoning behind this is that such persons have poorer sickness-absence records and tend to retire prematurely, thereby placing an unnecessary burden on the superannuation scheme. By the same token, smokers have an even greater reason for exclusion but the illogicality of this stance appears to elude such employers.

Occupational history

A crucial part of any examination for current and future use is the occupational history. Every medical student should be taught to take a full occupational history, for if they do not, sooner or later, an important aspect of a patient's history will be omitted and delay in diagnosis or treatment will result. The description 'civil servant' or 'retired' provides no useful information about the patient's job. More detailed information may indeed give the clue to the cause of the patient's complaint. At the very least, it will assist the physician in subsequently assessing the individual's ability to undertake his or her previous job with or without modification.

Ramazzini emphasized this point over 200 years ago when he noted in the introduction to his book *De Morbis Artificium* how important it was to ask the patient the nature of his work. It is, he says, 'concerned with exciting causes and should be particularly kept in mind when the patient belongs to the common people. In medical practice, attention is hardly ever paid to this matter though, for effective treatment, evidence of this sort has the utmost weight.'

The brief account should contain information on the following items:
• duration of job
• hours of work
• job title, with description of work done
• types of exposure including specific questions concerning dust, fumes, gases, liquids, temperature, noise, radiation, lighting, microbiological hazards, ergonomic and psychological factors.

Whilst all physicians should gauge the exact nature of the patient's present job, it behoves the occupational physician to take a full occupational history. Previous occupations may be the cause of the patient's current health problems, particularly for diseases of long latent period, such as cancer. In these circumstances, a detailed past history is vital and an example of such a review is included in Appendix 3.2, which the authors have also found useful in epidemiological studies. It is important to remember that the patient's pastime or hobbies may, in addition, cause diseases of occupation, e.g. the amateur boat builder who becomes allergic to isocyanates or the bird fancier who develops extrinsic allergic alveolitis. The occupational history is thus not only an essential of the pre-placement screen but is indispensable in any thorough medical assessment.

Post-sickness-absence examination
This is one of the most valuable routine medical assessments in industry. It provides the physician with an opportunity to match the job with the employee in the light of the recent illness. The assessment should always be undertaken and, indeed, might profitably be preceded by a review of the employee at four-weekly intervals *after* the *start* of his or her absence. The authors discovered a case of scrotal cancer this way, which led to a complete environmental reappraisal of the machine shop where the man worked and the establishment of regular screening procedures for all the other employees engaged in similar work.

A major reorganization of the certification of time off work attributable to sickness has been undertaken in the UK. The employer is now faced with establishing his/her own procedure for assessing the validity of such absences. Once an absence has been assessed as probably valid, the self-certification by the worker of the cause should be reviewed by the company's own medical adviser. Appropriately used, this responsibility will enhance the occupational health department's surveillance procedures over sickness absence.

Furthermore, new regulations promulgated in the UK in 1985 concern the Reporting of Injuries, Diseases and Dangerous Occurrences (RIDDOR). Certain diseases require notification to the HSE. The responsibility for such reporting rests with the employer. The first few years' experience of RIDDOR indicates considerable under-reporting. This finding has led the UK Health and Safety Executive to review its occupational ill health data collecting procedures.

Whether the physician needs to personally examine all return-to-work employees is debatable. Certainly they need to review all cases and insist on seeing those that fulfil the following criteria:
- absence longer than four weeks
- absence following an accident at work

- absence of more than two weeks following *any* accident
- absences attributable to vertigo, syncope, neoplastic, cardiovascular or neurological disease, infectious disease (food handlers), certain patterns (alcoholism, psychiatric disorders).

It is important that close links are established between the employee's general practitioner and the occupational health physician, as this will be invaluable in sorting out the difficult cases returning to work. Particular tact and skill are necessary in reaching an agreed plan of action where the general practitioner and the occupational health physician initially disagree over the correct timing of the return to work and on the suitability of the patient to resume his or her job in its original or modified form. Anticipation of such problems is preferable to a collision course over the correct management, particularly as, in the final analysis, the general practitioner is the employee's primary medical caretaker.

Periodic medicals for vulnerable groups
Seldom should time be devoted to medicals at the expense of post-sickness-absence reviews but there are cases where such assessments are useful and, for some workers, such as food handlers or drivers, they may be necessary or even mandatory. In addition medicals for young persons are still required in the coal mining industry.

Drivers
Public service and heavy goods vehicle drivers are statutorily required to be examined on appointment and periodically thereafter. London Regional Transport, for example, carries out these periodic reviews at the time of licence renewal following the 50th, 56th, 59th, 62nd and 64th birthdays and annually thereafter. Their French counterparts review annually after the start of employment. It is important to remember that, if such periodic reviews are undertaken and, particularly, if they are carried out more frequently than legally necessary, there should be an agreed and acceptable procedure within the company for cushioning the effect on an employee of an adverse medical report. Some maintenance of previous earnings, with or without an assurance about continued employment in another capacity, is essential if these reviews are to proceed with confidence and completeness through the full co-operation of the examiner and the examinee.

Specific age-groups
The young and the old may require specific medical review. Although the statutory examination of young persons entering employment has been

abolished in most industries (it rarely discovered anything other than tinea capitis, scabies or refractive errors), there is still a tendency to 'keep an eye on' such persons. This is more from the point of view of their psychological adjustment from school to work than for strictly physical reasons. By the same token, the approach of retirement, combined with a slowing and weakening of physical faculties, requires careful and considerate handling. The occupational health department is best suited to perform this delicate task. Pre-retirement consultation is also of great value.

Women
Women are frequently classified as 'vulnerable' but this is often on dubious grounds. Apart from some lesser muscle mass and physical strength, women are usually more resilient than men and they certainly live longer. Nevertheless, during reproductive age, they are vulnerable to gonadotoxic occupational hazards (a risk they share with men) and to fetotoxic agents (a risk uniquely theirs). In addition, women at work frequently remain the primary caretaker of the home and family. Their working day may, therefore, be 4–6 hours longer than their husbands'! Apart from these specific problems, women are just as able to cope with the average job as men. They are, however, frequently discriminated against on unscientific evidence. Nevertheless, EC regulations are being drafted to protect women—particularly those of reproductive age. The effectiveness of such measures—even their usefulness—remains to be seen.

Food handlers
Food handlers present a risk not so much to themselves as to the consumers of the products they process, make or serve. Applicants for food-handling jobs would normally be disqualified for such posts if they had
• inflammatory disease of the ear
• inflammatory disease of the eye
• persistent or recurring skin conditions subject to infection
• persistent upper or lower respiratory disease (e.g. bronchitis, tuberculosis)
• oral sepsis
• a history of typhoid or paratyphoid disease
• a history of intestinal disease in the absence of microbiological evidence of freedom from infection.

In addition, some companies are strict enough to bar nail biters and insist on reviewing the health of employees who have become unwell after recent travel outside Northern Europe or North America. Obviously, it is essential that all food handlers are medically reviewed prior to returning to work after absence attributable to sickness. However, the advent of the Food Safety Act 1990 will impose additional requirements on employers in the name of consumer protection. In particular, there will be a need for food hygiene training for relevant staff.

Indeed, the problems of training for food handlers are mirrored by the difficulties of health surveillance. Whilst canteen staff at the workplace may be relatively easy to keep under surveillance, many large retailing organizations employ staff at scattered 'outlets'. These premises could be anything from a large hotel employing 200 people to a public house employing a few 'casual' workers. One organization with such logistical difficulties has designed a series of health questionnaires with varying degrees of thoroughness commensurate with the degree of surveillance that is feasible. These form Appendices 3.3 and 3.4.

Executives

This is often considered to be a group suitable for routine periodic medical examinations. Notwithstanding the expense of replacing a senior manager prematurely retired through ill health, there is little justification for the multitude of such examinations done in the name of preventive medicine and which provide little benefit, apart from some perceived peace of mind on the part of the examinee. Multiphasic health screening programmes abound, though the evidence to date from carefully constructed controlled trials shows little prognostic value in the battery of tests to which such a manager is subjected.

Before embarking on an executive screening programme, it is worth applying the following criteria to any test considered of value:
• the condition for which the test is made should be an important health problem
• if found, it should have an accepted treatment
• facilities for diagnosis and treatment should be available
• the condition should have a recognizable pre- or early-symptomatic stage
• a suitable screening test should be available
• the screening test should be acceptable to the screened person
• the cost of case finding should be economical
• the case-finding procedure should be periodic rather than 'one-off'.

From the above, it should be clear that it is difficult to lay down specific guidelines for all occupations and all medical conditions. However, considerable thought should be applied to the drafting of a policy for the company regarding periodic medical examinations. An example of such guidelines for a large brewing organization is to be found in Appendix 3.4 Medical time has been reduced to a minimum and the occupational health nurse has been elevated to a position of crucial importance. The system has been in operation for ten years and has proved acceptable and workable to the semi-autonomous trading companies, each employing their own medical and nursing staff. That is not to say that the executive medical is a total waste of time. It can start (or cement) a useful working relationship and it can provide an opportunity for health counselling. Ideally, this health review should be specific to the executive's job requirements as for any other designated 'risk' groups in the industry.

Surveillance levels
The variation of medical surveillance, depending on the risk assessment of particular jobs, is summarized below.

Risk level	Records	Responsible person
Low: no occupational hazard	Basic	Worker/personnel officer
Low: possible occupational hazard	Basic + some medical*	Worker/nurse
Moderate	Routine screening*	Nurse/doctor
High	Full medical*	Doctor

* Where possible, workplace environmental data should be recorded, the degree of detail given being dependent upon the risk level. Such a system is incorporated in the COSHH regulations (Section 3.2).

Biological monitoring
The old medical joke about measuring the serum-rhubarb levels is close to reality these days, due to rapid advances in toxicological knowledge and analytical capability. Nevertheless, the truly professional biochemist will baulk at undertaking more than a limited range of biochemical analyses on human tissue or fluids, because of the serious and sometimes insurmountable obstacles of inter- and intra-laboratory variations in the results obtained. For example, in one study of blood-lead estimations of known concentration sent to a large number of participating and accredited European laboratories, the blood-lead results ranged from 100 μg/l to 1200 μg/l. The furore currently surrounding the recommendation that

'acceptable' occupational blood-lead levels be reduced from 800 to, perhaps, 600 μg/l is thus put in its true light.

Biological monitoring can be defined as the determination of exposure to and absorption of a workplace chemical by analysis of a biological sample, e.g. blood, urine. Again, as with physical examination, care must be taken to ensure that the test under consideration is undertaken for the right reasons and with a full knowledge of the errors which may be inherent in the procedure. Biological monitoring has expanded greatly in recent years. One result of this has been the appearance of Biological Exposure Indices (BEI). The following criteria are worth noting:
• biological monitoring is no substitute for environmental monitoring. The former assesses the individual risk, the latter assesses the need for engineering controls; both are necessary
• a good biological monitoring test need not necessarily correlate well with environmental levels, mainly because of human factors and differing routes of absorption
• the number of substances capable of reliable use is still small
• rapidly acting substances are usually unsuitable for biological monitoring
• the substance and/or metabolite to be measured must be present in some tissue or body fluid suitable for sampling
• valid and practical analytical methods for such measurement must be available
• the measurements must be correct
• the result should be interpretable in terms of health risk
• results often have greater value for the group examined than for the individual
• the problems of deciding on a 'suitable' concentration of a sensitizing agent must be considered.

Having decided that a test for substance X is appropriate, further questions arise:
• Which compound should be measured? The substance itself, a metabolite, or both?
• Which biological fluid or tissue is to be sampled?
• In relation to what period of exposure?
• How frequently should the sampling be done?

It is important to distinguish biological monitoring of *exposure*—an essentially preventive measure—from biological monitoring of *effect* (biological effect monitoring) an early detection device for toxicity. The former is frequently a measure of the ambient chemical following absorption—for example, the lead in blood concentration. The latter, in this case, could be blood zinc protoporphyrin or urinary δ-aminolaevulinic acid.

Table 3.1 contains some biological measurements where estimates are available of biological exposure indices.

In summary, it is important to differentiate three aspects of certain components of periodic health reviews:
• biological monitoring—the measurement and assessment of workplace agents or their metabolites either in tissues, secreta, excreta or any combination of these to evaluate exposure and health risk compared with an appropriate reference
• biological effect monitoring—the measurement and assessment of early biological effects, of which the relationship to health impairment has not yet been established, in exposed workers to evaluate exposure and/or risk compared with an appropriate reference
• health surveillance—the periodic medicophysiological examination of exposed workers with the objective of protecting from, and preventing occupationally related diseases (the detection of *established* disease is outside the scope of this definition).

Medical records
The quality and quantity of medical record-keeping will vary with the extent and sophistication of the occupational health service. Nevertheless, it is essential that a service provided by a nurse with a part-time doctor should at least maintain the following records:
1 Accident and dangerous occurrences book—frequently required by law.
2 Daily clinic attendance recorded by name, age, sex, department and reason for visit.
3 Treatment: type, site and reason.
4 Individual record cards, preferably one for each employee but certainly one for every person seen by a member of the occupational health staff. This will record treatments, clinic visits and results of routine examinations, as well as personal details, including unique identity numbers such as NHS or National Insurance numbers.
5 Periodic medical examination data. These should, preferably, allow space for pre-placement examination details and have continuation sheets for periodic assessments. They should be uniform for all personnel with or without distinguishing features for hazard areas, vulnerable groups or disease categories.

3.3 Periodic medical examinations

Table 3.1 Biological monitoring for selected chemicals

Chemical	Biological index	Normal value	Suggested maximum permissible value
Cadmium	Cadmium in urine	<2 µg/g creatinine	10 µg/g creatinine
	Cadmium in blood	<0.5 µg/100 ml	1 µg/100 ml
Lead	Lead in blood	<35 µg/100 ml	60 µg/100 ml (adult males)
			40 µg/100 ml (young females)
	Zinc protoporphyrin in blood	<2.5 µg/g haemoglobin	12.5 µg/g haemoglobin
Mercury	Total inorganic in blood	<2 µg/100 ml	<3 µg/100 ml
Benzene	Phenol in urine	<20 g/g creatinine	<45 mg/g creatinine (if TLV = 10 p.p.m.)
Toluene	Hippuric acid in urine	<1.5 g/g creatinine	2.5 g/g creatinine
1,1,1-Trichloroethane	Trichloroethanol and trichloroacetic acid in urine	—	50 mg/g creatinine
Carbon monoxide	Carboxyhaemoglobin in blood	<1%	5% (non-smokers)

---

6   Clinic attendance summaries by week, month and year for the main disease categories.

Such record systems are readily transferred and stored on computer. Such a computerized system *must,* however, have a device to restrict unauthorized access.

3.4   Occupational hygiene survey techniques

Monitoring the concentration of airborne substances in the workplace may be an essential part of the assessment required under the COSHH regulations. It certainly would be required to check whether an occupational exposure standard is being exceeded or, where a maximum exposure limit applies, whether the concentration is below this as far as is reasonably practicable. Regular monitoring is required by law for worker exposure to vinyl chloride and chromium plating vapour or mist. Regular monitoring would also be required to check the continued satisfactory operation of methods of controlling airborne contaminants such as local exhaust ventilation systems.

The COSHH Code of Practice describes monitoring as the periodic or continuous sampling of the airborne concentration of substances in the workers' breathing zone by means of personal sampling equipment, using valid and suitable occupational hygiene techniques. Personal sampling equipment has to be carried by workers whilst performing their normal duties and must be, of necessity, lightweight and socially acceptable.

For most airborne pollutants the sampling techniques involve drawing a known volume of contaminated air through a collection medium for subsequent analysis at the completion of the survey. The collection medium is placed in a holder which is hung as close to the nose of the worker as possible without interfering with his/her work. With such a technique the normal peaks and troughs of concentration that occur during the operations are not recorded, the result being a time-weighted average concentration over the period sampled. With most pollutants this is acceptable as the standard published is based on an eight-hour time-weighted average. However, for some substances a ceiling value is given which may not be exceeded. This requires either short-term (grab) sampling or a continuous indication of concentration, possibly coupled to an alarm warning device.

The published standards quote the concentration gravimetrically or volumetrically, that is, in either the units of mass of pollutant per unit volume of air, such as milligrams per cubic metre (mg/m^3), or volume per volume, such as parts of the pollutant per million parts of air (p.p.m.). The technique for sampling, therefore, is to collect a sample of the pollutant for as long as possible, usually the whole working shift, whilst carefully noting the flow rate of air passing through the sampler and the duration of sampling. The chemical analyst will provide the number of milligrams of sample collected, which is then divided by the total cubic metres of air sampled to give the gravimetric result. The volumetric result is either calculated from the gravimetric or displayed on a direct reading instrument.

Techniques of sampling are described in occupational hygiene text books and in some Health and Safety Executive publications (EH and MDHS series). For airborne particles such as dust or fibres, the collecting medium is a filter paper or membrane, usually 25 mm in diameter, carried in a holder and suitable for the size of particle to be collected. For airborne gases or vapours the medium is solid adsorbent such as activated charcoal held in a small tube or liquid carried in an impinger. The device for drawing the air through the collector is normally a small battery-powered pump having a flow rate to suit the quantity of sample required but normally between 0.002 and 4 litres per minute.

Uncertainties of sampling
It is essential to obtain a sample of the worker's exposure that truly represents a typical exposure and here is where the skill of the occupational hygienist is vital. It is important to realize that to obtain a realistic picture of the health hazards of a working environment no single measurement will suffice. This is because workplace pollution rarely occurs evenly spread in concentration or intensity over the whole workplace or over the whole working period. In the case of emission of a gas or particles of dust the concentration is greatest at the point of emission but it may fluctuate as the process progresses. As the pollutant moves away its behaviour and dispersion will depend upon the air currents occurring in the room and therefore will vary with the movement of people, machinery and both mechanically and naturally induced air currents. Even one shift may not be typical of others in the same place due to the variability and cyclic nature of the processes.

Therefore surveys must be planned within the resources available to obtain the best possible information on the hazards as they affect individual workers and the workforce as a whole. Invariably, the more measurements that are taken over the longest periods of time, the more valuable the results will be. If important decisions are to be made or expensive equipment purchased based upon the results of measurements, then they must be taken as accurately and as scientifically as possible and preferably by a professional occupational hygienist.

3.5   Occupational hygiene standards for airborne contaminants

Standard-setting authorities
Many countries set their own occupational exposure standards but the most active are the Federal Republic of Germany, USSR, Holland, USA, UK and most recently the EC. A comparison of standards can be seen in a publication by the International Labour Office (ILO), Geneva, entitled Occupational Exposure Limits for Airborne Toxic Substances.

United States
Two organizations have produced standards which have influenced the rest of the world—the American Conference of Government Industrial Hygienists (ACGIH), Cincinnati, and the Occupational Safety and Health Administration (OSHA), Washington.

ACGIH Threshold limit values (TLVs) have been produced over many years and copied by many other countries worldwide including, until recently, the United Kingdom. Their publication, updated annually, gives a list of over 500 chemicals and shows against each chemical an eight-hour time-weighted average (TWA) and a short-term exposure limit (STEL) and they indicate which chemicals are likely to be absorbed through the skin. Methods for calculating standards for mixed exposures are given.

OSHA Permissible exposure limits (PELs) are enforced by their factory inspectors and each is contained in a four-page document giving advice on first aid, medical surveillance, measurement in the workplace, methods of control, personal protective equipment and emergency actions. Over 200 substances are covered.

United Kingdom
The Health and Safety Executive (HSE) publish standards in their Guidance Note EH40/year, e.g. EH40/1991. The format is similar to the ACGIH TLVs but the standards are in two categories as defined by the COSHH Regulations—maximum exposure levels (MELs) and occupational exposure standards (OESs). These are defined as follows.

MELs At present (1991) 30 MELs are published and for those substances exposure should be controlled as far as is reasonably practicable below that level. Under the COSHH Regulations it is an offence to exceed them; they are listed in Schedule 1 of the Regulations and Table 1 of EH40. They

---

are produced by a tripartite committee (employers, employees and scientists) called the Advisory Committee on Toxic Substances (ACTS). MELs apply to a substance for which there is some concern, for example, carcinogens or those which could cause irreversible damage such as lead. The respiratory and skin sensitizers, for example, formaldehyde or isocyanates, also fall into this category. A maximum exposure limit should not be regarded as a safe level, that is, it will not protect everyone exposed at work and for that reason it is necessary to get as far below it as is reasonably practicable. New MELs are added annually.

OESs Approximately 500 of these standards exist and they should not be exceeded, but where they are, the reasons should be identified and the situation remedied as soon as reasonably practicable. Many of them have the same values as the ACGIH TLVs.

EH 40/year This is revised and reprinted annually. In addition to MELs and OESs it gives lists of changes from the previous year as well as substances to be reviewed. It gives advice for single substances, with the exception of some complex mixtures of compounds such as white spirit, rubber fume and welding fume, but, where mixed exposures are involved such as paint solvents, advice is given.

Mixed exposures

The ways in which the constituent substances of a mixture interact vary considerably. Some mixtures involve substances that act on different body tissues or organs, or by different toxicological mechanisms; these are regarded as being 'independent'. Other mixtures will include substances that act on the same organs or by similar mechanisms, so that the effects reinforce each other and the substances are said to be 'additive'. Sometimes the effect is considerably greater than the sum of the individual effects and these are known as 'synergistic'.

Independent effects As the substances act independently it is only necessary to ensure compliance with each of the individual exposure limits.

Additive effects Exposure should be assessed using the formula:

$$\frac{C_1}{L_1} + \frac{C_2}{L_2} + \frac{C_3}{L_3} + \ldots \text{ must not exceed 1}$$

where C is the airborne concentration and L is the exposure limit.

Synergistic effects Very little is known about these but they are regarded as being uncommon. Advice from an occupational health toxicologist should be sought if there is any suspicion.

Dusts
Dusts in substantial concentrations, as mentioned in Regulation 2, are defined in EH40 as 10 mg/m^3 eight-hour TWA for 'total inhalable dust' and 5 mg/m^3 eight-hour TWA for 'respirable dust'. Certain specific dusts are listed.

Difficulties with standard setting
• Reliable standards frequently cannot be set as there is a shortage of information on human experience and a dearth of animal studies.
• It is difficult to define what is a 'safe working level'.
• People's susceptibility varies as does their exposure dose due to variations in workload and duration fo exposure.
• A set standard should be accompanied by advice on how to measure the exposure.
• Socio-economic factors play a part in setting of standards.
• Little is known about the combination of physical factors such as heat, noise and stress on the uptake of substances into the body.
• Eight-hour exposure assumes 16 hours in clean air to rest and recuperate, which is not always the case.

European standards
Committees have been set up to commission and review criteria on which EC member states can base their decisions for setting standards. Some 28 criteria will soon be published and about 30 will be produced per year.

Actions when no standards exist
There are many substances for which no standards are set. This does not mean that these are safe; it means they have not been tested or considered. If guidance is required the following actions could be taken
• Look for an overseas standard, e.g. Germany, Holland, USA, or consult the ILO list.
• Find a company using that substance and see what they do.
• Find a substance that is chemically similar for which there is a published standard.
• Set up an investigation within the company and other users to see if any health effects have been noticed against known airborne exposures.
• In the case of therapeutic drugs, calculate an airborne standard based

upon a daily inhalation quantity being some fraction of the published oral or parenteral therapeutic dose levels.

3.6   Health survey design—the epidemiological approach

Introduction
Occupational health practitioners are responsible for the well-being of populations of people at work. Groups of the workforce have certain workplace exposure characteristics in common; thus any system of health status review must take population characteristics into account as it cannot rely on single patient observations which is the basis of clinical medical training. This population-based approach is the epidemiological method.

Definition of epidemiology
Epidemiology is defined as the study of the distribution and determinants of disease frequency in human populations.

Purpose
Some of the uses of epidemiology are outlined in Section 1.5. It is important to note, however, that epidemiology should not be considered the sole preserve of the academic. Indeed, the opposite is frequently true. If the occupational health practitioner has a question concerning the health of the workforce in relation to a workplace exposure, then epidemiology is frequently the way forward. Much of the present knowledge about the effects of occupational exposures has come from epidermiological studies.

The stages in an epidemiological study
There are ten basic stages in the design of an epidemiological study
- formulate the question
- review the literature
- decide on the survey design
- write the protocol
- assess the feasibility
- pilot the procedures—questionnaire, examination, investigations
- conduct the study
- analyse the results
- publicize the findings
- consider the need for future preventive action on exposure.

3 Evaluating workplace hazards

3.6 Health survey design—the epidemiological approach

Survey design options
Despite the confusing terminology in many text books, the options are limited. In essence, the question is related to how best to tackle the following situation:

Putative exposure(s) $\rightarrow$ observed health effect(s)

The options are:

cross-sectional (prevalence)
or

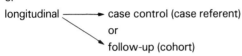

A cross-sectional study gives a snap-shot view of events. It cannot relate exposure to effect. This requires a longitudinal component, which means either a case control or a follow-up study. These are illustrated below.

Case control studies

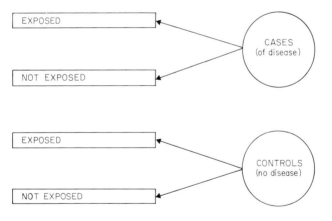

The controls are similar to the cases except (ideally) for the occurrence of the disease. The analysis consists of comparing the exposed/not exposed *ratio* for the cases with that of the controls.

Follow-up studies
The analysis here allows disease *rates* to be calculated by exposure category.

Note the direction of the arrows linking exposure with health effects.

3.6   Health survey design—the epidemiological approach

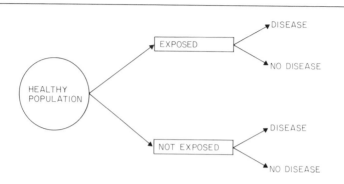

A case control study starts with the disease and works backwards (retrospectively) towards the exposures. Such studies, although highly suitable for rare diseases and relatively quick and easy to perform, are prone to bias of various types

- selection of cases or controls
- comparability of the cases and controls (excluding the disease factor)
- recall information on exposures
- accuracy of diagnosis of health effects, etc.

Follow-up studies are free from many of these biases if conducted in a truly forward (prospective) manner. Whilst they are suitable for rare exposures, they are time-consuming and thus expensive.

The object of the exercise is to decide whether exposure A ($\pm$ others) causes health effect B ($\mp$ others). The danger is that such an association could be spurious even if statistically significant. This is because of limitations in the method and/or confounders. A confounder is a factor which can independently affect the exposure and the outcome. A good example is age. Where possible such confounders should be identified and, if possible, allowed for by such procedures as matching (in the design stage) or stratifying the results by the confounder (in the analysis stage).

Cause or association?
In the end, if a relationship has been found between cause and effect, then the answers to a number of questions can help to distinguish true cause from apparent association

- Is the disease specific to a special group of workers?
- Is it much more common than in other groups of people?
- Has the association been described by others in other relevant populations?

- Does the exposure *always* precede effect by a biologically plausible time interval?
- Is there a dose–response effect?
- Are there corroborative laboratory (*in vitro*) or animal (*in vivo*) data?
- Whatever the statistical significance, is the result biologically plausible?

Shortcomings in the epidemiological method
Epidemiology is not an exact science and some of the major flaws are listed below
- a healthy worker effect—the comparison group has a different general health status compared with the cases
- poor response rate
- high turnover of study populations—selecting in (or out)
- latency between exposure and effect longer than study period
- insufficient evidence of differing effects by differing exposures
- poor quality of health effects data
- poor quality of exposure data
- multiple exposures
- no effect of exposure noted—does this mean true negative result or merely poor/small study (*non-positive result*!)?

Appraising an epidemiological study
Even if the occupational health practitioner never intends to undertake an epidemiological study, s/he should be able to evaluate a published study and assess its worth.

    The following points should be considered in any publication
- question clearly formulated
- appropriate study design
- good-quality health effects data
- good-quality exposure data
- valid population choice for cases/control
- high response rate/good sampling strategy
- confounders considered and allowed for
- attempts made to reduce bias
- population large enough to detect an effect if present
- correct statistical techniques
- estimates of risk include measures of variability, e.g. confidence intervals
- cause/association issues addressed
- non-positive/negative study result reviewed
- effect of results on current knowledge assessed.

Note: No epidemiological study is perfect but the method is frequently the only option in assessing potential human health effects. The procedures and processes are governed by the practicalities of the situation—availability of subjects, exposure estimates, etc.—but the aim is to seek the answer to an exposure–effect question. If such a relationship is found, measures must follow which diminish the likelihood of further health effects on this population or on subsequent generations of workers. Epidemiology is an aspect of preventive medicine and not just an intellectual exercise for an idle academic with a penchant for 'number-crunching'.

Appendix 3.1   Pre-employment medical forms

HEALTH RECORD

PART I   IN CONFIDENCE
For completion by Personnel Department

Region ....   Company ....

Name ....   Date of joining ....

Address ....

....

Tel. no. ....   Date of birth ....

National Insurance no. ....

Name and address of family doctor ....

....   Tel no. ....

1st change ....   2nd change ....

....

Tel. no. ....   Tel. no. ....

3 Evaluating workplace hazards

Appendix 3.1 Pre-employment medical forms

For completion by the applicant
A   What job are you going to do for the Company? ....
B   Please list your jobs, starting with the last one and working back to school.

	Dates From	To	Job	Company	Leave clear for comments from the occupational health staff
1	....	....	....	....	....
2	....	....	....	....	....
3	....	....	....	....	....
4	....	....	....	....	....
5	....	....	....	....	....
6	....	....	....	....	....
7	....	....	....	....	....
8	....	....	....	....	....
9	....	....	....	....	....
10	....	....	....	....	....

PART II
For completion by the applicant
Please answer the following questions, answering Yes/No as appropriate. If answer is 'yes' to any question, please give detail in 'Remarks' column.

Question	Yes	No	Remarks
1 Have you, *ever in your life,* including childhood, had any of the following:			
Allergies, e.g. hayfever, drugs, etc.	....	....	....
Blackouts or epilepsy	....	....	....
Heart trouble	....	....	....
Raised blood pressure	....	....	....
Tuberculosis	....	....	....
Diabetes	....	....	....
Asthma, bronchitis or pneumonia	....	....	....
Nervous disorders, 'nerves' or breakdown	....	....	....
Dermatitis or other skin diseases	....	....	....
Ear infections	....	....	....
Back trouble—causing time off work	....	....	....
Foot or knee trouble	....	....	....
Varicose veins	....	....	....
Rupture	....	....	....
Any other illness, e.g. jaundice, stomach ulcer, malaria, dysentery, anaemia, etc.	....	....	....

3   Evaluating workplace hazards

Appendix 3.1   Pre-employment medical forms

HEALTH RECORD (Continued)

2 Is your eyesight satisfactory, wearing glasses
if necessary? ....

3 Are you at present having any injections,
pills, tablets or medicines prescribed by a doctor? ....

4 Have you ever suffered from any accident or
disease requiring hospital admission or treatment? ....

5 Have you stayed away from work (or school)
in the last year? If so, for how long? ....

6 Have you had a chest X-ray in the past year? ....
If 'yes' was it normal? ....

7 How many children have you got? .... Were they normal births? Yes/No.
If abnormal, please give details ....

Have you/your spouse had any miscarriages?   Yes/No.
If yes, please give details ....
....

PART III
For completion by Medical Department

1 Family history

Mother ....

Father ....

Siblings ....

....

Children ....

2 Physical examination

Height .... cm

Weight .... kg   BP $\dfrac{\text{Systolic}}{\text{Diastolic}}$ ——— mmHg

Smoking/day .... cigarettes .... cigars .... tobacco (oz)

Alcohol/day .... pints of beer .... glasses of wine .... measures of spirits

3   Evaluating workplace hazards

Appendix 3.1   Pre-employment medical forms

HEALTH RECORD (Continued)

Vision	*Right*	*Left*
Distant	Corrected/Uncorrected	Corrected/Uncorrected
Near	Corrected/Uncorrected	Corrected/Uncorrected
Colour		

Urine

Glucose ....   Protein ....

Other ....

3 Menstruation

Menarche ...... years   Duration of bleed ...... days

Interval between bleeds ...... days   Menopause ...... years

4 Current medications ....

....

Conclusions..

....

Fit ....

Refer to doctor ....

---

3   Evaluating workplace hazards

Appendix 3.1   Pre-employment medical forms

HEALTH RECORD (Continued)

PART IV: INITIAL MEDICAL EXAMINATION
For completion by Medical Adviser

	Normal	Abnormal	If abnormal, describe
1 General physical appearance			
2 Cardiovascular system			
Pulse			
Heart			
Arteries			
Veins			
ECG			
3 Respiratory system			
Breath sounds			
Chest movement			
Chest X-ray			
4 Gastrointestinal system			
Teeth			
Mouth			
Abdomen			
Hernial orifices			
Rectal examination			
5 Central nervous system			
Pupils			
Fundi			
Cranials			
Reflexes			
Plantars			
Power			
Sensation			
Co-ordination			
6 Genito urinary system			
7 Glands			
Lymphatic			
Thyroid			
Breasts			
8 Skin (including scars)			
9 Ear, nose and throat			
Ear			
Nose			
Throat			
10 Skeletal system			
11 Psyche			
12 Blood tests			
Haematology			
Electrolytes			
Liver function			
Other			
13 Audiometry			

3 Evaluating workplace hazards

Appendix 3.1   Pre-employment medical forms

Conclusions and recommendations

I certify that in my opinion the candidate is (circle where appropriate):

A   Fit for all types of employment.

B   Unit for employment.

C   Temporarily unfit. Re-examine in .... months' time.

D   Fit for all types of employment, but with the following conditions (see below).

E   Fit for the proposed employment *only*.

Remarks : ....

....

Category ....   Signature of doctor ....

Resurvey in .... months   Name in print ....

                                                 Date ....

PART V: PERIODIC MEDICAL REVIEW

To be completed by nurse and/or doctor

Examination	Date									
Any diseases, accidents, medical treatment since last examination	......	......	......	......	......	......	......	......	......	......
Drugs: current medication	......	......	......	......	......	......	......	......	......	......
Height	......	......	......	......	......	......	......	......	......	......
Weight	......	......	......	......	......	......	......	......	......	......
Blood pressure	......	......	......	......	......	......	......	......	......	......
Smoking	......	......	......	......	......	......	......	......	......	......
Alcohol	......	......	......	......	......	......	......	......	......	......
Vision	......	......	......	......	......	......	......	......	......	......
Distant: left	......	......	......	......	......	......	......	......	......	......
right	......	......	......	......	......	......	......	......	......	......
Near: left	......	......	......	......	......	......	......	......	......	......
right	......	......	......	......	......	......	......	......	......	......
General physical appearance	......	......	......	......	......	......	......	......	......	......
Cardiovascular system	......	......	......	......	......	......	......	......	......	......
Respiratory system	......	......	......	......	......	......	......	......	......	......
Gastrointestinal system	......	......	......	......	......	......	......	......	......	......
Central nervous system	......	......	......	......	......	......	......	......	......	......
Genito-urinary system	......	......	......	......	......	......	......	......	......	......
Glands	......	......	......	......	......	......	......	......	......	......
Skin	......	......	......	......	......	......	......	......	......	......
Ears, nose and throat	......	......	......	......	......	......	......	......	......	......
Skeletal system	......	......	......	......	......	......	......	......	......	......
Psyche	......	......	......	......	......	......	......	......	......	......
ECG	......	......	......	......	......	......	......	......	......	......
Chest X-ray	......	......	......	......	......	......	......	......	......	......

3 Evaluating workplace hazards

Appendix 3.1 Pre-employment medical forms

HEALTH RECORD (Continued)

Examination	Date									
Blood tests	.....	.....	.....	.....	.....	.....	.....	.....	.....	.....
Haematology	.....	.....	.....	.....	.....	.....	.....	.....	.....	.....
Electrolytes	.....	.....	.....	.....	.....	.....	.....	.....	.....	.....
Liver function	.....	.....	.....	.....	.....	.....	.....	.....	.....	.....
Other	.....	.....	.....	.....	.....	.....	.....	.....	.....	.....
Audiometry	.....	.....	.....	.....	.....	.....	.....	.....	.....	.....
Recommendation (Category A–E)	.....	.....	.....	.....	.....	.....	.....	.....	.....	.....
Date of next examination	.....	.....	.....	.....	.....	.....	.....	.....	.....	.....
Signature of examiner	.....	.....	.....	.....	.....	.....	.....	.....	.....	.....

Medical notes for non-routine consultation

Date	Medical notes
....	....
....	....
....	....
....	....
....	....
....	....
....	....
....	....
....	....
....	....
....	....

Medical report

Date of Examination ....   Surname ....

Occupation ....   First Name(s) ....

Found to be (circle where appropriate)

A   Fit for all types of employment.

B   Unfit for employment.

C   Temporarily unfit. Re-examine in .... months' time.

D   Fit for all types of employment, but with the following conditions (see below).

E   Fit for the proposed employment *only*.

Remarks: ....

....

....

....

Signed ....

Appendix 3.2 Occupational history pro-forma

We need information on your past work experience, and substances you worked with and were exposed to. Start with your present job (if currently not working, your last job) and go back over all the jobs you have held for as long as 3 months. Include moonlighting jobs. Please provide as much information as you can remember. Include all military service.

What is your usual occupation? ....

Dates of job (years)	Name of company (employer)	What did the company make or do?	City/ County	Job title	What did the work involve?	Was the work considered dangerous? If yes, why?	Was protective equipment or clothing available? If yes, what was it?	Did you inhale chemical solvents, dust or other fumes? If yes, list if known	Did chemical oils, dusts, etc., get on your skin/ clothes? If yes, list if known

Appendix 3.3   Medical questionnaires

UNIT STAFF MEDICAL QUESTIONNAIRE

Surnames: ....Mr/Mrs/Ms/Miss   Forename(s): ....

Unit address: ....

Position: .... Registered disabled: Yes/No   Number: ....

		Yes	No
1	Do you have or have you ever suffered from:		
A	Typhoid fever		
	Paratyphoid fever		
	Fits or blackouts since the age of five		
B	Recurring ear infections		
	Recurring skin trouble affecting hands, arms and face		
	Recurring disabilities affecting standing, walking, lifting or use of the hands		
2	At present, are you suffering from:		
A	a discharging ear		
B	boils, styes or septic fingers		
C	a cough with phlegm		
D	abdominal pain, diarrhoea or flu		
E	skin trouble affecting the hands, arms or face		

Answers to these questions are accurate to the best of my knowledge.
I acknowledge that failure to disclose information may require reassessment of my fitness and would lead to termination of employment.

Signature: .... (Prospective employee)

Signature: .... (Interviewer)

Date: ....

Appendix 3.3   Medical questionnaires

UNIT MANAGEMENT MEDICAL QUESTIONNAIRE

This completed health record is confidential to the Medical Department

Surname: .... Forename(s): .... Date of Birth: ....

Unit Address: ....

Position: .... Registered Disabled: Yes/No   Number: ....

This form is to be completed by the applicant. Both partners in a couple should complete a separate form. Please answer the following questions, ticking the appropriate Yes/No box. If answer 'Yes' to any question please give details in 'Remarks' column.

		Yes	No	Remarks
1	Have you, *ever in your life*, including childhood, had any of the following:			
	Allergies, e.g. hayfever, drugs, etc.			
	Blackouts or epilepsy			
	Heart trouble			
	Raised blood pressure			
	Tuberculosis			
	Diabetes			
	Asthma, bronchitis or pneumonia			
	Nervous disorders, 'nerves' or breakdown			
	Dermatitis or other skin disorders			
	Skin infections			
	Back trouble—causing time off work			
	Varicose veins			
	Rupture			
	Fainting attacks			
	Giddiness			
	Recurring stomach trouble			
	Recurring bowel trouble			
2	Have you any disabilities affecting:			
	Standing			
	Walking			
	Stair climbing			
	Lifting			
	Use of hands			
	Working at heights			
	Ability to drive motor vehicle			

Appendix 3.3   Medical questionnaires

UNIT MANAGEMENT MEDICAL QUESTIONNAIRE (Continued)

	Yes	No	Remarks
3 Have you ever had:			
Typhoid fever			
Paratyphoid fever			
Ear trouble			
A running ear			
Chest trouble with cough and phlegm			
4 At present are you suffering from:			
A cough with phlegm			
Acne, boils, styes or septic finger			
Diarrhoea, abdominal pain, fever			
A running ear			
5 Have you visited your dentist within the last 6 months?			
If dental treatment is necessary, are you willing to visit your dentist?			
6 Is your eyesight satisfactory, wearing glasses if necessary?			
7 Are you at present having any injections, or taking pills, tablets or medicines?			
8 Have you ever suffered from any accident or disease requiring hospital admission or operation?			
9 Have you stayed away from work (or school) in the last year? If so, for how long?			
10 Have you had a chest X-ray in the past five years? If yes, was it normal?			

11 Height: cm	Weight: kg (*in usual clothes, without shoes*)
Please give any further details of health problems or disabilities, not covered above which could result in absence from work	

The answers to these questions are accurate to the best of my knowledge. I acknowledge that failure to disclose information may require a reassessment of my fitness and could lead to termination of employment.

Signed: ....   Date: ....

Appendix 3.3   Medical questionnaires

FOREIGN TRAVEL MEDICAL QUESTIONNAIRE

To be completed for all Unit Staff after travel outside EC and North America

Employee's name: ....

Address: ....

....

....

Company: ....

---

1   State countries visited:

A .... Length of stay .... Dates ....

B .... Length of stay .... Dates ....

C .... Length of stay .... Dates ....

2

A Have you been suffering from sickness, diarrhoea or
bowel disorders?   Yes/No

B Have you been suffering from any infectious condition
of the skin, nose, throat or eyes?   Yes/No

If yes to **A/B**:
C Are you still showing any symptoms from your illness
which might indicate an incomplete recovery?   Yes/No

3

A Have you any flu-like symptoms?   Yes/No

B Have you been in contact with anyone who has typhoid,
paratyphoid or cholera?   Yes/No

Signature of employee: ....   Date: ....

Appendix 3.4 Routine medical examinations

Examinee	Examiner	Examination and health review procedure	Periodicity
Grades 1–12	Nurse	Review of health questionnaire and BP, visual acuity, height, weight, urinalysis, refer to doctor (?)	Pre-placement, then as required
Grades 13 +	Doctor	Review of health questionnaire, general medical examination, possibly including chest X-ray and ECG	Pre-placement, then as required. Routine biennial review (in-house or others)
Drivers HGV* FLT Salesmen Others driving more than 40 000 miles per annum for Company	Doctor	Review of health questionnaire. General medical examination, with special emphasis on: Eyesight—including visual fields and fundoscopy Ears—including hearing and vestibular function Cardiovascular system—including BP, heart disease (past and current) Nervous system—including psyche, epilepsy, power, co-ordination and sensation Musculoskeletal system—including mobility of limbs and vertebral column	Pre-placement, then 5-yearly[†] to 50 years 2-yearly to 60 years Yearly to 65 years
Food handlers and canteen staff[‡]	Nurse[§] Doctor[§]	Review of health questionnaire. General review of personal hygiene and habits, with special emphasis on: Infection of eyes, ears, nose, throat, skin, teeth Past or current gastrointestinal pathology	Pre-placement. Post any sickness absence Post any leave in countries other than North Europe or North America
Managed house staff	Nurse[§] Doctor[§]	Weekly visits to canteen As for canteen staff supplemented by reports from Area Managers	Pre-placement, then 5-yearly

Appendix 3.4 (Continued)

Special examinations

Noise hazard area	To be established
Laboratory staff	Annual mercury in urine estimations (for mercury-exposed group)
Dicolyte handlers	Biennial chest X-ray
VDU operators	Pre-placement visual acuity and field tests
Post-sickness absence	Review of reasons for absence by doctor and nurse at least monthly, at best weekly
Post-accident-absence	Review of reasons for absence by doctor and nurse at least fortnightly, at best weekly

* See Raffle A. (ed.) Medical Aspects of Fitness to Drive, 4th edn. Medical Commission on Accident Prevention, London.
† 3-yearly for HGV.
§ See Employment Medical Advisory Service (November 1972) Medical requirements for fork lift operators. Notes of Guidance, EMAS, London; and Health and Safety Executive (1979) Lift Trucks, HMSO, London.
‡ Whether contract workers or company employees.
NB: union approval may be required for some or all of these categories of driver listed above

4 Occupational diseases

4.1   Historical perspective, 83

4.2   Notifiable and prescribed diseases, 85

4.3   General principles of toxicology, 88

4.4   Target organs, 91

       Appendix 4.1   Prescribed diseases in the UK, 111

4.1   Historical perspective

People were subject to hazards in their daily life long before the Industrial Revolution and the advent of industrial workplaces. The vagaries of climate and food supply, not to mention the menace of the sabre-toothed tiger, provided prehistoric man with quite enough risks to health. True occupational hazards must have arrived, however, by the time of the Stone Age, as the knapping of flints produces small clouds of silica dust. Whilst our ancestors are unlikely to have lived long enough to die of silicosis, the fashioning of iron tools and the development of mining and smelting certainly increased the dangers for those so engaged.

Indeed, mining was recognized in ancient Egyptian times as being so hazardous that the job was reserved for slaves and criminals. It was not until Agricola (1494–1555) and Paracelsus (1493–1541) formally recorded these risks to employed medieval artisans that any real attention was paid to their plight. By the sixteenth century, mining had become a skilled occupation and Agricola not only described the hazards but prescribed some remedies, such as improvements to ventilation and mine shaft design, which were necessary to diminish the staggering death rate in the mines of Joachimstal and Schneeberg. The same Carpathian mountains are still mined today but, instead of silver (used to make the Joachimstaler = taler = dollar), the ore extracted is uranium. This might explain the rampant 'consumption' noted by Agricola, as it was probably lung cancer caused by radioactive gases and dusts.

Nevertheless, the first general and authoritative treatise on diseases related to occupations was written by the physician to the D'Este family in Modena, Bernardino Ramazzini (1633–1714). His book *De Morbis Artificium* is still unparalleled as a source of classic descriptions of many occupational disease, ranging from those of cesspit workers to those of the mirror silverers of Murano. His work was largely unread until the Industrial Revolution in Britain brought occupational diseases to the attention of large numbers of people. Child labour and the atrocious working conditions in the cotton mills of Lancashire shocked many late Georgians and early Victorians and the first factory legislation was pushed through by philanthropic factory owners such as Robert Owen, Michael Sadler, Anthony Ashley Cooper (Earl of Shaftesbury) and Robert Peel, despite some stiff opposition.

The first Act of 1802 was greatly weakened by amendments in Parliament but it started the process of legislation protecting workers which culminated in the Health and Safety at Work Act, etc. (1974). In between these dates, successive Acts reduced the hours of work,

4 Occupational diseases

4.1 Historical perspective

particularly of women and young children, and the Act of 1833 established the factory inspectorate. Four inspectors were appointed to cover the whole country. Eleven years later, the inspectors were given the additional power to appoint certifying surgeons in each district to decide on the age of children. The advent of birth certification in 1836 eventually made that role redundant but an embryonic industrial medical service had been born. Later Acts gave these surgeons other duties, including the investigation of industrial accidents and the certification of fitness for work.

By the turn of the twentieth century the toxic effect of certain materials in widespread use in industry was sufficiently well recognized in the UK to warrant their notification. This provided the power to investigate incidents of disease, with a view to prevention. The first agents so notified in 1895 were lead, phosphorus, arsenic and anthrax. The list later extended to some 16 diseases notifiable by law.

The flood of notifications that emanated from the print factories, match works, smelters and slaughterhouses, in the nineteenth century, necessitated the appointment in 1898 of the first medical inspector of factories, Sir Thomas Legge (1863–1932). Legge is famous for many things, not least the furore over his resignation in 1919 due to the Government's refusal to ratify an international convention prohibiting the use of white lead for the inside painting of buildings. His aphorisms, though sounding a little paternalistic to late twentieth century ears, are still worth citing:

1 Unless and until the employer has done everything (and everything is a lot) the workman can do little to protect himself.

2 If you bring an influence to bear external to the workman (i.e. one over which he can exercise no control), you will be successful and if you do not, or cannot, you will *not* be successful.

3 Practically all lead poisoning is due to inhalation.

4 All workers should be told something of the hazards of the materials they work with—if they find out for themselves it may cost them their lives.

5 Influences, useful up to a point but not completely effective when not external but dependent on the will or whim of workers, include respirators, goggles, gloves, etc.

Today, the task of providing occupational health services for all workers is still an unattained ideal even in developed countries. Furthermore, although many of the older occupational diseases are controlled or are eliminated, new ones continue to surface.

4.2   Notifiable and prescribed diseases

No one knows how much ill-health is attributable to occupation. The official figures can provide total sickness-absence, industrial accidents and prescribed disease (Fig. 4.1) but this is, almost certainly, a gross underestimation of occupationally related disorders. (Recent changes in the claims procedures for prescribed diseases prevent valid current comparison but the general proportional relationship holds true for the 1990s as it did for the 1980s.)

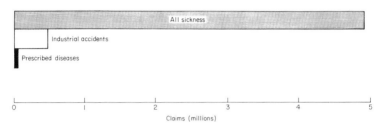

Fig. 4.1 Male sickness claims in the UK, 1978–9 (Department of Health and Social Security 1981).

The diseases that used to be notifiable by law in the UK were
• lead poisoning (specified industries)
• cadmium poisoning
• beryllium poisoning
• phosphorous poisoning
• manganese poisoning
• mercurial poisoning
• arsenic poisoning
• carbon disulphide poisoning
• aniline poisoning
• chronic benzene poisoning
• toxic jaundice
• toxic anaemia
• decompression sickness
• anthrax
• epitheliomatous ulceration
• chrome poisoning.
    The notifiable diseases list has been superseded by the RIDDOR list. There are also a large number of occupational diseases prescribed by the Department of Social Security for the purposes of compensating

4   Occupational diseases

4.2   Notifiable and prescribed diseases

individuals shown to have contracted them. The prescribed diseases list is
similar but not identical to those conditions reportable by employers under
the Reporting of Diseases and Dangerous Occurrences Regulations
(RIDDOR) 1985.

The consideration of diseases for prescription under the industrial
injuries scheme is the responsibility of the Industrial Injuries Advisory
Council (IIAC), established specifically for this purpose. The Council may
accept or reject the arguments for prescription and thereby advise the
Secretary of State accordingly. The main points it considers in assessing a
disease are whether:

1   It ought to be treated, having regard to its causes and incidence and
any other relevant considerations, as a risk of their (workers') occupations
and not a risk common to all persons.

2   It is such that, in the absence of special circumstances, the attribution
of particular cases to the nature of the employment can be established or
presumed with reasonable certainty.

There were 63 prescribed diseases in the UK by mid-1991. The
schedule of prescribed diseases is listed in Appendix 4.1.

Table 4.1 provides details of the ten most commonly compensated
occupational diseases in the United Kingdom. It should be noted, however,
that, for non-respiratory diseases diagnosed after October 1986, the
claimant had to be assessed at more than 14% disabled to receive regular
benefit. This has resulted in a dramatic fall in the published figures for
disease which may be compensated.

It is, however, known that prescription figures grossly underestimate
the true toll of occupation-related disease. For example, in Finland where

Table 4.1 Top ten prescribed industrial diseases: new cases 1984–1989

Disease	1984	1985	1986	1987	1988	1989
Occupational deafness	1468	1492	1179	1381	1515	1506
Vibration white finger	N/A	3	641	1366	1673	1056
Pneumoconiosis	577	702	747	652	562	661
Diffuse mesothelioma	201	245	305	399	479	441
Tenosynovitis	337	390	619	376	322	294
Dermatitis	611	619	785	464	368	285
Occupational asthma	144	165	184	213	180	214
Bilateral pleural thickening	N/A	61	111	115	114	125
'Beat' conditions	131	180	220	57	171	112
Lung cancer in asbestos workers	N/A	8	34	55	59	54

Source: HSC 1990

vital statistics recording is accurate and compensatable disease schemes are more generous, the occupational deafness figures are similar to the UK, whilst the 'repetitive strain disorders' and skin diseases figures are ten times larger. Given that the Finnish working population is ten times *smaller* than in the UK, the underestimation of the toll of occupational disease in Britain would be up to 100-fold!

So far as occupationally related mortality is concerned, the Decennial Supplements of the Office of Population, Censuses and Surveys are a rich source of 'hypothesis generation'. Figures 4.2 and 4.3 are two of the myriad graphs that can be extracted from the microfiche tables. Figure 4.2 shows the influence of socio-economic status on all causes of mortality, which is more marked in men than in 'single women'. Similarly, the pot-pourri of occupational groups depicted in Fig. 4.3 illustrates the varying mortality experience. Caution is, however, necessary. Too much should not be read into these figures as bias in both the numerator and the denominator is not inconsiderable. Yet, as a source reference for considering occupational factors in mortality, the OPCS publications are without parallel.

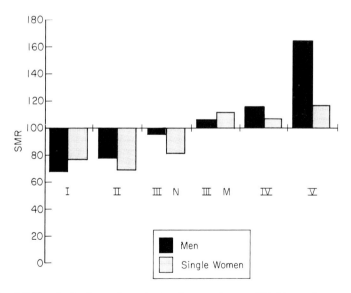

Fig. 4.2 Standardized mortality ratios for all causes for men (20–64 years) and single women (20–59 years) by socio-economic group. I = Professional, II = Intermediate, III N = Skilled non-manual, III M = Skilled manual, IV = Semi skilled, and V = Unskilled.

4.3   General principles of toxicology

Occupation

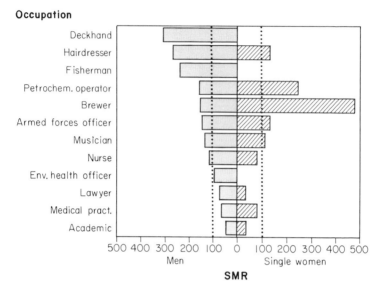

Fig. 4.3 Some occupational standardized mortality ratios (SMR) for men and single women—UK. Source: *Dec. Suppl. Occ. Mort.* 1979–80, 1982–83 OPCS 1986.

4.3   General principles of toxicology

Toxicology concerns the potential of chemicals to produce adverse effects in the body. Whilst the occupational health practitioner cannot be expected to be a toxicologist, it behoves him or her to ensure that s/he is fully conversant with the toxicological aspects of the materials in the workplaces for which s/he has responsibility. Although modern toxicology hopes to advance by what it learns from *in vitro* and *in vivo* laboratory based experiments, it is not uncommon, even today, for the human effects of exposure to new chemicals to be discovered at the cost of worker health. This tragic state of affairs was the rule rather than the exception a century ago.

The first principle of toxicology is that all substances can kill if given by an inappropriate route and in inappropriate quantities—even water or sodium chloride. In industrial terms, therefore, the *hazard* of a given substance is related to
• its absolute toxicity
• its physico-chemical properties
• circumstances of use—concentration, length of exposure, body area exposed.

4.3   General principles of toxicology

For example, the cyanide radical (CN^-) is highly toxic to biological enzyme systems; yet hydrocyanic acid (HCN) and sodium cyanide (NaCN) have differing degrees of lethality. HCN is a gas and NaCN is a white crystalline powder which, though capable of dissolution in water, does not release HCN until it reacts with, say, gastric hydrochloric acid.

The routes of entry of toxic materials may be
• inhalation—the most common route, industrially
• ingestion—unusual (but may follow ingestion of inhaled particles returned via the ciliary escalator)
• skin absorption—more common than is realized if the material is fat-soluble.

Once in contact with the body, the toxic material may have varying effects:
• local—irritant to the skin, eye or respiratory tract or allergic to (usually) the skin or respiratory tract
• systemic—inherent toxicity, metabolite toxicity or both.

It is possible, therefore, for a toxic material to produce local effects at the point of contact as well as distant effects during its peregrinations through the body. The most common target organs are the
• skin
• lungs
• liver
• nervous system
• bone marrow
• kidneys.

The skin is the commonest site for effects of exposure to toxic chemicals. Contact dermatitis is one of the commonest occupational diseases.

The lungs are obviously at risk as an organ of note that 'meets' inhaled toxins. The liver and kidneys are important organs of metabolism and excretion and are, therefore, likely to be confronted by the toxic substance or its metabolite(s), if absorption has taken place. Many industrial poisons are fat-soluble and, therefore, find their way to organs with a high fat content, such as the brain, spinal cord and bone marrow. In addition, their toxic effects may be prolonged, due to temporary storage in body fat.

It is naive to consider that the liver is well equipped to 'detoxify' industrial poisons. The truth of the matter is that the liver will apply its standard metabolic processes to such foreign material. These are either oxidation, reduction or hydrolysis (phase I) or synthesis, such as conjugation with glucuronic acid (phase II). Phase I reactions tend to increase the polarity of the metabolized material; phase II reactions tend to increase its acidity. Both measures improve water solubility and, thereby, aid urinary excretion. Less common routes of excretion include the bile, the skin and the lungs.

Not everyone reacts equally to a given dose of a toxic material. The factors that may be of significance in determining toxic effect include

- age
- sex
- ethnic grouping
- genetic background
- endocrine status
- atopic state
- nutrition
- fatigue
- coexistent disease (and its treatment)
- coexistent exposure to other synergistic (or antagonist) chemicals— including prescribed medications
- previous exposure to the toxic agent.

There is no single suitable classification of toxic substances. Various authorities class them chemically, physically or physiologically. A frequently quoted measure of relative *acute* toxicity is the LD_{50}, that is the dose of the material capable of killing half a population of exposed laboratory animals shown in the following table.

Toxicity rating	Description	LD_{50} (wt/kg single oral dose to rats)
1	Extremely toxic	1 mg or less
2	Highly toxic	1–50 mg
3	Moderately toxic	50–500 mg
4	Slightly toxic	0.5–5 g
5	Practically non-toxic	5–15 g
6	Relatively harmless	15 g or more

Such an impressive grading system does not allow for a slow or delayed death and, perhaps more seriously, assumes that humans react to these materials as though they were rats. Yet there is no all-purpose and appropriate toxicity grading device and the LD_{50} is often quoted. It is important, therefore, to be aware of its limitations and to note that a not inconsiderable number of occupational exposure limits are based on such tenuous evidence—flimsy justification indeed for workplace control criteria.

The need for a more appropriate agreed scale of toxicity—particularly with the burgeoning of toxicity testing systems which avoid the use of laboratory animals—is urgent.

4.4   Target organs

Introduction
In this section, consideration is given to the organs of the body and the way they respond to insult and assault from occupationally related agents. Detailed toxicological and pathological information on each agent cited is found in Chapter 5. However, it is the organ's response to injury which will usually herald the onset of ill health. This is what brings the worker to the attention of physicians. Chronic plumbism, for example, may present itself in a variety of system dysfunctions and the physician will have to unravel the differential diagnosis: it is a rare event indeed for a patient to complain of overexposure to inorganic lead!

The target organ systems included in this section are the

• respiratory system
• central and peripheral nervous system
• genitourinary system
• cardiovascular system
• skin
• liver
• reproductive system

Respiratory system

Structure
The upper and lower respiratory tracts are particularly vulnerable to occupationally related noxious agents. Over 80% of these agents gain access to the body through the respiratory system. The effects of such exposure may also be felt in other organ systems but the brunt of the damage frequently falls on the air passages and lungs.

The system is composed of several anatomically discrete sections

• the mouth, nasal sinuses, pharynx and larynx
• the trachea, main and segmental bronchi
• the bronchioli
• the alveoli
• the alveolar–capillary barrier.

4.4   Target organs

The repeated branching of the airways from tracheal bifurcation to alveoli has the effect of greatly increasing the surface area of respiratory mucosa, whilst reducing the rate of air flow. Thus, the 300 million alveoli offer a surface area of some 70 m^2 for gas exchange but no alveolus exceeds 0.1 mm in diameter. The thickness of the alveolar epithelial wall, together with the endothelial cell layer of the pulmonary capillaries is, in health, rarely greater than 0.001 mm and constitute the blood–gas barrier.

Function
The main function of the lungs is to supply oxygen for uptake by the pulmonary capillaries and to provide the means for removing carbon dioxide diffusing in the opposite direction. The successful achievement of this gas exchange requires three main system functions
• ventilation
• gas transfer
• blood-gas transport.
Lung function tests can be similarly grouped.
  Ventilatory function is commonly measured by a variety of portable instruments, the most common of which are the peak flow meter and the vitallograph. Despite recent developments in microcomputer analysis, which have been of great benefit in providing rapid digital and graphic read-outs of ventilatory function, the main indices are
• forced expiratory volume in 1 s (FEV_1)
• forced mid-expiratory flow rate (FMF)
• forced vital capacity (FVC)
• the FEV/FVC ratio
• flow/volume patterns: compliance, elasticity.
Figure 4.4 illustrates these indices.
  Gas transfer and transport measures require less portable apparatus and are a means of assessing
• ventilation/perfusion ratios—mainly involving radioactive gases
• gas diffusion—usually transfer of carbon monoxide, Tl_{co}
• red cell gas uptake and transport—usually blood-gas levels and pH.
  Lung function is altered by a number of non-occupational factors and tables of values of some of the variables are available. They include
• age
• sex
• lung size
• ethnic grouping
• height

4   Occupational diseases

4.4   Target organs

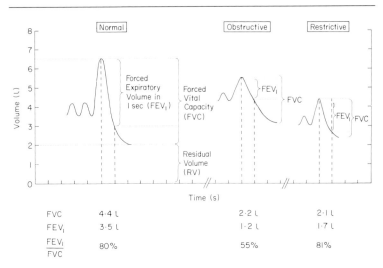

	Normal	Obstructive	Restrictive
FVC	4·4 l	2·2 l	2·1 l
FEV₁	3·5 l	1·2 l	1·7 l
FEV₁/FVC	80%	55%	81%

Fig. 4.4 Spirograms to illustrate the differences in forced expiratory volume in 1 s (FEV$_1$) and forced vital capacity (FVC) in health, airways obstruction and restrictive defects (such as diffuse lung fibrosis or severe spinal deformity). Inspiration is upwards, expiration downwards. Although the vital capacity is reduced in both obstructive and restrictive lung disease, the proportion expired in 1 s shows considerable differences. (From Waldron and Harrington (1980) *Occupational Hygiene* p. 70. Blackwell Scientific Publication, Oxford.)

* smoking habits
* exercise tolerance
* observer error
* instrument error
* diurnal variation
* ambient temperature.

   Finally, the chest radiograph is a valuable tool for assessing lung exposure to occupationally related dusts. The purpose is twofold:

1   Clinical—to establish diagnosis and prognosis and as a guide to treatment.

2   Epidemiological—to assess prevalence and progression of disease both in populations and individual members of populations.

   To this end, a standard classification of radiographs has been internationally agreed for use in assessing pneumoconiosis. This ILO U/C classification, as it is called, requires a chest radiograph to be taken in a standard fashion, read and classified to a standard format. The main features of this format are listed in Table 4.2.

4   Occupational diseases

4.4   Target organs

Table 4.2 ILO U/C classification of pneumoconiosis

Feature	Classification
No pneumoconiosis	0
Pneumoconiosis	
Rounded small opacities	
Profusion	1, 2, 3
Type	p, q(m), r(n)
Extent	Zones
Irregular small opacities	
Profusion	1, 2, 3
Type	s, t, u
Extent	Zones
Large opacities	
Size	A, B, C
Type	Well-defined/ill-defined
Pleural thickening	By site
Diaphragmatic outline	Ill-defined: right, left
Cardiac outline	Ill-defined 1, 2, 3
Pleural calcifications	By site and extent

Profusion relates to the area of spread across the lung fields. *Type* relates to the size of the opacities. *Extent* relates to the zones affected. The 1, 2, 3 profusion groups can be extended to a 12 point system: 0/−, 0/0, 0/1, 1/0, 1/1, 1/2, 2/1, 2/2, 2/3, 3/2, 3/3, 3/4.

Occupational lung disorders

Harmful effects to the lung produced by noxious agents can be grouped into six categories:

*   acute inflammation
*   asthma
*   byssinosis
*   pneumoconiosis
*   extrinsic allergic alveolitis
*   malignancy (see Section 8.6.)

Acute inflammation This is primarily caused by the irritant gases and fumes described earlier. Their solubility determines whether their effects are most noticeable on the upper or lower respiratory tracts. They include

*   ammonia
*   chlorine
*   sulphur dioxide
*   nitric oxide
*   phosgene

94

- fluorine
- ozone.

Occupationally related asthma This may be caused by a variety of dusts, either causing an immediate reaction or a late (non-immediate) type. The former may develop within minutes of exposure, the latter may take 4–24 hours to develop (usually 4–8 hours). It is in this latter group that the suspicion of a work-related cause may be missed, as the effect of the allergenic dust or fume may not be noticed until the evening of the day of exposure. Some people experience a combination of immediate and late effects. A suggested classification for two types of occupational asthma appears in Table 4.3.

Occupational asthma has been variously estimated as causing 2–15% of all asthmas but it is important to realize that atopy is *not* a predisposing state for more than a few of the known allergenic dusts. Although a good occupational and medical history may settle the diagnosis, some cases are exceedingly difficult to unravel. In these cases, it may be necessary to resort to bronchial challenge testing, which, in competent hands, is a relatively safe procedure. Skin prick testing and serology may also be of value.

Treatment involves removal of the sensitized subject from exposure (in some cases this may necessitate a change of job) and non-specific treatment for the asthmatic symptoms.

Prevention involves several factors
- efficient and rigorous hygiene control in the workplace
- substitution with less allergenic materials
- personal respiratory protection for the worker

Table 4.3 Occupational asthma—possibly two distinct types

Agent	Examples	Atopy	Exacerbating factors
Large molecular weight protein antigen	Rodent urine Shellfish protein Flour/grain Mite faeces	Yes	Cigarette smoke +
Low molecular weight chemicals	TDI Acid anhydrides Platinum salts Plicatic acid (woods) Abietic acid (colophony)	No	Cigarette smoke + +

• identification of workers at risk—not an easy task where 30% of the population have atopic characteristics and atopy is not necessarily a prognostic feature of occupational asthma
• periodic medical examinations with pre- and post-shift ventilatory capacity measurements.

The list of potential occupational lung allergens is now extremely long. A few of the more important ones are
• grain, flour, hops, tobacco dust
• beetles, locusts, cockroaches, grain mites
• laboratory animals, e.g. rats, mice
• avian feathers
• amoebae ('humidifier fever')
• fungi
• various woods, including cedars, boxwood
• various metals and their salts, including platinum, chromium, nickel, vanadium
• formalin
• ethylene diamine
• chloramine T
• phthalic anhydride (epoxy resin hardener)
• colophony (soldering flux)
• certain dyes
• drugs including penicillin, tetracycline and methyldopa
• enzymes including *B. subtilis* and pancreatic extracts
• isocyanates.

Byssinosis This is considered by many authorities to be a type of occupational asthma. The condition is, however, broader and more complex than this. In susceptible individuals, exposure to the dusts of cotton, sisal, hemp or flax can produce acute dyspnoea, with cough and reversible airways obstruction. It is first noticed on the first day of the working week and then subsides. Later, with continued exposure, symptoms recur on subsequent days of the week until even weekends and holidays are not free of symptoms.

The effects are greatest where the dust concentrations are highest and are more noticeable with coarser cotton. This has led to the suggestion that the condition is (in part) due to organic contaminants of the cotton boll such as bracts and even microbiological agents such as *E. coli*. Smoking exacerbates the condition and, although irreversible airways obstruction may eventually supervene and even kill the patient, no specific pathological proccesses have been indentified at post-mortem in the lungs. Chest radiography is unhelpful and treatment is symptomatic.

Recent studies of Ulster flax workers and Lancashire cotton workers have cast doubt on the ability of this condition to influence mortality rates. Byssinosis as described today is probably a mixture of conditions ranging from true asthma to exacerbated chronic bronchitis.

Pneumoconioses The term 'pneumoconiosis' literally means 'dusty lungs'. For practical purposes, pneumoconiosis is usually restricted to those conditions which cause a permanent alteration in lung architecture following the inhalation of mineral dusts. These dusts include
• silica (or quartz)
• coal
• asbestos.
The clinical and radiographic features of silicosis, coal workers' pneumoconiosis and asbestosis are summarized in Fig. 4.5.

Silicosis occurs following the inhalation of 'free' silica and is most common amongst workers involved in quarrying, mining and tunnelling through quartz-bearing rock, for example, gold mining. It is also a hazard of
• abrasives
• sand blasting
• glass manufacture
• stone cutting and dressing
• foundry work
• ceramic manufacture.
   Silicosis may be of four main types:
• nodular, with hyaline and collagenous lesions in the lungs
• mixed dust fibrosis, with irregular, stellate fibrotic lung lesions
• diatomite—a picture similar to fibrosing alveolitis and usually attributed to diatomaceous earth
• 'acute'—a rapidly developing alveolar lipoproteinosis with fibrosing alveolitis.

The fibrotic lung lesions frequently calcify and there is a progressive restrictive lung disease leading to cor pulmonale. Pulmonary tuberculosis used to be a common complication.

Recent epidemiological studies have suggested that silica may be a lung carcinogen. There certainly seems to be an increased risk of lung cancer in *silicotics* and some authorities suggest an additive risk of silica dust exposure and cigarette smoking. At present, the question is unresolved.

Coal dust produces a somewhat different picture and, frequently, has

4.4   Target organs

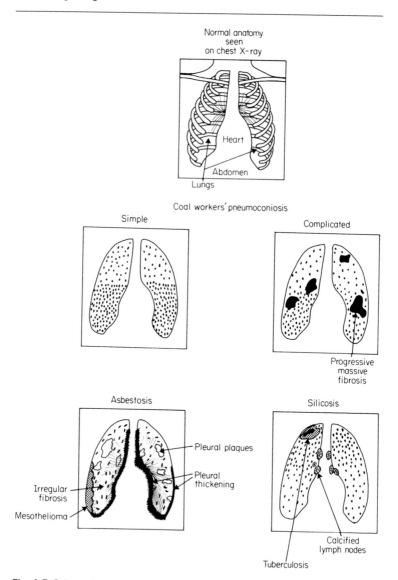

Fig. 4.5 Schematic representation of the chest X-ray appearances of certain dust diseases.

less severe sequelae and is less aggressively fibrotic. Indeed, the distinction between 'simple' and 'complicated' coal pneumoconiosis is often marked, with only a minority of the former progressing to the latter. What factors predispose to this serious turn of events are unknown.

4.4   Target organs

In the UK, the prevalence of *all* categories of coal pneumoconiosis has
fallen to less than 5%, with a prevalence of complicated (progressive
massive fibrosis) of 0.4%. The probability of developing ILO category 2 or
3 simple pneumoconiosis, over a working life-time in British mines, now
stands at between 2 and 12%, if dust levels are maintained at
concentrations below the current occupational exposure limits (3 mg m^{-3}
for operations where the average quartz level exceeds 0.45 mg m^{-3}, to 7
mg m^{-3} for long wall, coal-face operations). Indeed, most miners
nowadays have little to show radiographically for their work and, as long
as the lung mottling seen remains simple, the worker is usually
symptomless. Problems (and symptoms) start if and when the discrete
rounded opacities seen on chest X-rays coalesce, break down or become
more fibrotic. Serious pulmonary dysfunction then supervenes and death is
commonly due to cor pulmonale.

Asbestos fibres produce more irregular and more florid fibrotic changes
in the lung. The disease is thought to be progressive and death from
restrictive lung disease is not uncommon. Clinically, the disease may be
first manifest by dyspnoea, non-productive cough, finger clubbing or
weight loss. The presence of persistent chest pain frequently heralds a
recognized complication of asbestos exposure, namely malignant disease
— usually a bronchogenic carcinoma, occasionally a pleural mesothelioma.

For completeness, it is worth noting that some dusts seem to produce
disturbingly florid lung mottling on chest radiographs but with little or no
evidence of clinical effect or progression. These so-called benign
pneumoconioses are associated with the dusts of
• barium—barytosis
• tin—stannosis
• antimony, zirconium and the rare earths
• iron—siderosis (though haematite mining is associated with lung cancer,
thought to be due to radon gas in the mines).

Extrinsic allergic alveolitis
As noted earlier, the inhalation of organic materials can give rise to asthma.
However, other organic dusts produce an alveolitis, with the resultant
lowering of gas transfer across the blood–gas barrier. Most of the agents
capable of producing this effect are fungal spores and the most common
clinical condition in Britain is farmer's lung. These diseases frequently have
acute, influenza-like episodes which, if exposure is continued, lead to sub-
acute and chronic pulmonary fibrotic disease. The various types of
extrinsic allergic alveolitis are summarized in Table 4.4.

4.4   Target organs

Table 4.4 Types of extrinsic allergic alveolitis

Type	Exposure to	Allergen
Farmer's lung	Mouldy hay	*Micropolyspora faeni,* *Thermoactinomyces vulgaris*
Bagassosis	Mouldy sugar cane	*Thermoactinomyces sacchari*
Suberosis	Mouldy cork	*Penicillium frequentants*
Bird fancier's lung	Droppings and feathers	Avian protein
Malt worker's lung	Mouldy barley	*Aspergillus clavatus*
Air conditioner disease	Dust or mist	*T. vulgaris, T. thapophilus* and amoebae (various)
Cheese worker's lung	Mould dust	*P. casei*
Wheat weevil lung	Mouldy grain or flour	*Sitophilus granarius*
Animal handler's lung	Dander, dried rodent urine	Serum and urine proteins
Pituitary snuff taker's lung	Therapeutic snuff	Pig or ox protein

Nervous system

The basic unit of the nervous system is the neurone, which has four components as follows, with the nerve impulse passing from the first to the last:

• the dendrites
• the cell body itself
• the axon
• the synaptic terminal.

The axon is one long nerve fibre with or without myelin sheathing, which speeds nerve conduction. In the normal resting state, the axonal membrane has a resting potential of about -85 mV; this *outside* positive charge is maintained by an active sodium pump mechanism. A nerve impulse is a wave of depolarization and repolarization, which runs along the nerve fibre as the membrane permeability to sodium increases, allowing a rapid reversal of polarity followed by a recovery period as 'normal' permeability is restored. Conduction along the axon is all or nothing, but along the dendrites it is incremental. At the synapse, the electrical energy is transformed into chemical energy by the release of neurotransmitter such as acetyl choline. These neurotransmitters may be excitatory or inhibitory.

Occupationally related disorders of peripheral nerves
These may be motor or sensory nerve effects, commonly both. Sensory disturbances are usually distal; motor dysfunction may be proximal or distal. Most toxic substances cause axonal degeneration, usually by a mechanism which is largely unknown. The damage may be limited and

reversible or severe and permanent, depending on the agent, the dose and the duration of exposure. In the case of *n*-hexane and methylbutylketone, the effect seems to be due to a common metabolite, 2,5-hexanedione and leads to giant axonal swelling in the proximal parts of the axon, with peripheral 'dying-back'. Apart from the clinical features of peripheral neuropathy, subclinical effects may be detected by nerve conduction studies, which can be useful in the early detection of effect in workers exposed to known neurotoxins.

The most common neurotoxins are
- organophosphate pesticides
- carbamate pesticides
- triorthocresyl phosphate
- *n*-hexane
- methylbutylketone
- acrylamide and/or dimethylaminoproprionitrile (DMAPN)
- carbon disulphide
- mercury compounds (inorganic and organic)
- lead and its compounds (inorganic)
- arsenic
- thallium
- antimony.

Occupationally related disorders of the central nervous system (CNS)
Disordered brain function produces a consistent clinical picture, despite the complexity of the organ affected. The primary change is one of disordered consciousness, varying from mild disorientation through to profound coma. Organic psychoses are frequently worst at night and when the patient is fatigued. These effects may be interspersed with periods of normal mentation. In additon, visual hallucinations may occur, with or without memory disturbance, ideas of reference, paranoia, anxiety and, later, apathy.

Occupational toxins of these sort are frequently organic solvents or the heavy metals and include
- arsenic
- lead (including epilepsy)
- manganese (including Parkinsonism)
- mercury
- carbon disulphide (including Parkinsonism)
- organic lead compounds
- chlorinated hydrocarbons, pesticides such as dieldrin (including epilepsy)

4.4 Target organs

- carbon monoxide
- halothane
- methylene chloride
- trichloroethylene
- perchlorethylene
- toluene
- benzene
- methyl chloroform
- styrene
- white spirit
- possibly compressed air.

Screening for CNS effects is not easy and involves a battery of behavioural tests, including cognitive and perceptual psychomotor assessments. Recent international meetings have shown that an agreed battery of psychological tests can be formulated. Examples of such test procedure are available in the larger texts cited at this end of this book (Rom, Zenz).

What remains to be resolved is whether chronic low dose exposure to various organic solvents can cause organic psychosis. The 'Danish printers syndrome'—much vaunted in the early 1980s—has not been corroborated in American or British studies. The best of these studies are, however, not without a measurable effect. It appears that some tests of higher cerebral function do show a decremental change following long occupational exposure. The clinical and epidemiological signigicance of such results remain to be evaluated.

Alcohol abuse may exacerbate the effects of the occupational neurotoxin but these effects are, nevertheless, relatively rare. In normal clinical practice, diabetes is a more likely diagnosis for the patient with a neuropathy.

Genitourinary system

The kidney plays a crucial role in the excretory and detoxification mechanisms of the body. When a toxic substance is absorbed, the liver will frequently alter its chemical structure. Although it is naive to imagine that the liver 'knows' how to detoxify such foreign material, it is nevertheless a fact that the liver frequently succeeds in increasing the toxin's polarity and/or acidity. Both these changes will render the material more water-soluble and, hence, more readily excretable through the kidney. Some toxic substances reach the kidney unchanged; others reach the kidney in the form of a more toxic metabolite; yet others are able to cause damage by either being sequestered in the renal cortex (e.g.

cadmium) or by being present in the bladder long enough to cause malignant change (e.g. some of the aromatic amines such as 2-napthylamine and its metabolites).

Common forms of disordered function resulting from occupationally related substances include tubular dysfunction leading to aminoaciduria, proteinuria and glycosuria (e.g. mercury) or acute renal failure due to tubular necrosis (e.g. thallium), hypovolaemic shock (arsenic) or tubular blockage by crystalluria (oxalic acid). Cortical necrosis, whilst a rare cause of acute renal failure, occurs more commonly with the toxic nephropathies than with other causes of kidney damage. Furthermore, severe liver damage, due to, say, organic solvents, may induce a complication of renal failure and it is worth remembering that drug-induced hypersensitivity renal damage can occur, not only in those taking the drug therapeutically but also in those making the drug occupationally. The nephrotic syndrome can be occupationally induced if the proteinuria is of sufficient severity. This can occur following exposure to mercury, gold and bismuth.

Renal tract malignancy of occupational origin is an important condition, primarily affecting that part of the genitourinary system in contact with the agent for longest and at the highest concentration, namely the bladder.

Although the prostate possesses the curious ability to concentrate (and excrete) heavy metals, little incontrovertible evidence exists of occupationally induced prostatic disease in workers exposed to heavy metals. The putative link between cadium exposure and prostatic cancer has not been corroborated by more recent, careful studies.

A short list of the more important occupational factors in nephrotoxicity is given in Table 4.5.

Cardiovascular system

The cardiovascular system is not in the 'front line' when it comes to onslaught from noxious materials encountered in the workplace. Nevertheless, work may be a risk factor in the pathogenesis of cardiovascular disease, if only as a factor in stress-related illness. The classical risk factors for cardiovascular disease include

- age
- sex
- weight
- race
- cigarettes
- blood pressure
- serum cholesterol concentrations and diet

4 Occupational diseases

4.4 Target organs

Table 4.5 Occupational factors with significant nephrotoxic effect

Inorganic	Organic	Miscellaneous
Arsenic	Aniline	Antimicrobials
Bismuth	Carbon tetrachloride	Cantharides
Boron	Chloroform	Chlorinated
Cadmium	Dimethyl sulphate	hydrocarbon
Gold	Dioxan	insecticides
Iron salts	Ethylene glycol	Electric shock
(in overdose)	EDTA	Fungi
Lead	Methoxyfluorane	Horse serum
Mercury	Methyl alcohol	Trauma
Phosphorus	Methyl chloride	X-ray contrast
Potassium chlorate	Paraquat	media
Thallium	Pentachlorophenol	
Uranium	Phenol	
	Toluene	
	Turpentine	

- exercise levels
- personality
- stress
- oral contraceptives
- hardness of water (?)
- family history
- medical history (e.g. diabetes).

Hazards directly emanating from the workplace tend to have a secondary effect on the cardiovascular system. For example, lead can cause nephropathy, which in turn can cause hypertension. Cadmium causes emphysema, which can lead to cor pulmonale; cobalt can cause cardiomyopathy and arsenic is thought to be a factor in peripheral vascular disease.

Certain organic solvents are thought capable of inducing cardiac arrhythmias and vinyl chloride has been linked with Raynaud's phenomenon. Methylene chloride produces carbon monoxide as a metabolite, whilst carbon disulphide seems to possess a direct atherogenic potential. Finally, certain gases can cause hypoxia, directly or at a cellular level, and physical factors, such as vibration, can induce vasospastic disease in the small arteries of the hand.

Thus, the list of occupationally related factors which can cause, induce or exacerbate cardiovascular disease is not inconsiderable and includes

- lead
- arsenic

4   Occupational diseases

4.4   Target organs

- cadmium
- cobalt
- antimony
- manganese
- mercury
- 1,1,1-trichloroethane
- trichloroethylene
- chloroform
- carbon tetrachloride
- halothane
- fluorocarbons
- vinyl chloride
- glyceryl trinitrate and trinitrotoluene
- carbon disulphide
- methylene chloride
- carbon monoxide
- temperature
- vibration
- low atmosphere pressures
- electric shock.

Skin

The skin consists of two basic elements: an outer epidermis, which acts as a protective armour and is non-wettable, and a dermis, which provides the inherent strength of the skin, largely through its collagen content.

The waterproofing capability of the epidermis is, in occupational health terms, a potential problem, as the greasy surface aids absorption of fat-soluble materials and is, thus, a ready route of entry for many organic chemicals.

Skin disease may be characterized by discrete elevated lesions, patchy rashes which bear a limited geographic resemblance to the area of external assault and distinct localized irritation which, usually, is a faithful replica of the area of injury.

Occupational dermatoses may be broadly divided into two groups:

1   Primary irritant contact dermatitis.
2   Allergic contact dermatitis.

Primary irritant contact dermatitis
Nearly three-quarters of all occupational dermatoses are of this type. The irritants produce a direct effect on the skin with which they come into

contact and the effect will be more dependent on the dose and duration of exposure than on any inherent response emanating from the individual. For example, concentrated sulphuric acid splashed on to the face of anybody will produce a skin reaction. Soap and water are more variable but these seemingly harmless materials can cause irritation in non-allergic subjects. The degree of effect depends on such factors as

- skin dryness
- sweating
- pigmentation
- integrity of the epidermis (e.g. trauma)
- presence of hair
- presence of dirt
- concurrent or pre-existent skin disease
- environmental factors such as temperature, humidity, friction.

Furthermore, some chemicals, such as arsenicals, mercurials and chromium salts, are capable of combining with skin protein and causing ulceration. Cutting oils and chlorinated napthylenes can block sebaceous ducts and cause acne.

Allergic contact dermatitis
Sensitizing eczemas account for 15–20% of all occupational dermatoses. The response is usually specific to one agent but may be delayed for a week or more after contact. The initial sensitizing episode may need to be of several hours' duration, even though subsequent reactions can be provoked by the most transient exposures.

The mechanism of the response is a delayed hypersensitivity reaction. The allergen (hapten) combines with protein in the epidermis and is engulfed in skin macrophages and transported in the lymphatics. At the regional lymph nodes, circulating antibody is produced which is then 'ready' to react locally with any further contact with the hapten–protein complex. The acute effect is erythema, eruption, vesiculation, oozing and desquamation. In a chronic form, this leads to thickened fissured skin.

The main contact allergens are

- dichromates
- epoxy resins
- rubber accelerators and antioxidants
- germicidal agents such as hexachlorophene
- dyes such as paraphenylenediamine
- topical anaesthetics such as procaine

4.4   Target organs

- formadehyde
- nickel and its salts
- mercury and its salts
- cobalt and its salts

   Many of these can cause occupationally related dermatoses and, in addition, the following are particularly potent:

- antibiotics
- colophony
- vinyl and acrylic resins
- plants: primula, daffodil, chrysanthemum
- West African hardwoods
- pharmaceuticals such as chlorothiazide, phenothiazines and tolbutamide.

   The industries with greatest likelihood of causing occupationally related skin disease are

- leather goods
- food processing and packing
- adhesive and sealant
- boat building and repair
- abrasive products
- agriculture and horticulture.

The liver

The liver is the largest visceral organ. Its mass of parenchymal cells, portal tracts and abundant blood supply is witness to its crucial role as the body's main metabolic factory. The functional unit resides in the lobule and consists of a group of sinusoids running between a terminal portal tract and a few terminal hepatic venules. The liver cells nearest the hepatic venules differ from those near the portal tracts, in that the former receive blood lower in oxygen content. These centribolular hepatocytes are therefore more vulnerable to toxic (and anoxic) conditions than the periportal cells.

   Disordered hepatic function has two main aspects. Firstly, the hepatocytes may be damaged or, secondly, the transport mechanism to, through or from the hepatocytes may be blocked. Both dyfunctions, will, sooner or later, lead to jaundice. The liver's remarkable capacity for hepatocyte regeneration following assault is not matched by its ability to reproduce, faithfully, the basic liver architecture. Thus, major hepatic damage can lead to florid cellular regeneration in a disrupted lobular pattern.

4.4   Target organs

The effects of liver injury can be viewed thus:

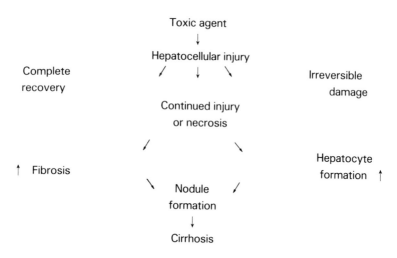

Pre-existent liver disease may enhance the effects of a new toxic onslaught and the liver is, thus, particularly vulnerable to the effects of organic solvents if already reeling under the effects of regular and excessive dosing with ethyl alcohol. Furthermore, it is worth noting that hepatic enzyme induction may alter the liver's ability to handle certain toxins and the organ itself is occasionally subject to hypersensitivity reaction. The effects of hepatotoxins of occupational origin may be subclassified as in Table 4.6.

Reproductive system

It was not until the last two decades that serious attention was paid to the possibility that occupational factors could disrupt the normal functioning of the reproductive system. Anecdotal evidence suggests that women working with lead in the early years of the century became sterile, but the recent furore over visual display units and pregnancy outcome has heightened the awareness that reproduction could be disrupted by work exposure.

Now, considerable attention is being paid to the possibility that gonadal function, for example, may be altered or even arrested following occupational exposure to certain toxins. Animal studies have indicated a wide range of chemicals that can disrupt reproductive capacity. For the male, these include steroidal hormones, alkylating agents, antimetabolites, diuretics, psychopharmacological agents, anaesthetics, oral

4.4   Target organs

Table 4.6 Effects of occupationally related hepatotoxins

Effect	Substance	
	Organic	Inorganic
Centrilobular necrosis	Acrylonitrile	Antimony
	Carbon tetrachloride	Arsenic
	Chlorinated hydrocarbon	Boranes
	insecticides	Phosphorus (yellow)
	Chlorinated napthylenes	Selenium
	Dimethyl hydrazine	Thallium
	Dimethyl nitrosamine	
	Dinitrophenol	
	Ethyl alcohol	
	Halothane	
	Methyl chloride	
	Nitrobenzene	
	Phenol	
	Polychlorinated diphenyls	
	Tetrachloroethylene	
	1,1,1-Trichloroethane	
	Trichloroethylene	
	Trinitrotoluene	
	Toluene	
	Vinyl chloride	
Hepatic effect	Halothane	
	Viral hepatitis	
	Leptospirosis	
Cholestatic cholangiolytic	Methylene dianiline	
	Organic arsenicals	
	Toluene diamine	
	4,4'-diaminodiphenyl	
	methane	

hypoglycaemics, ethyl alcohol, heavy metals, insecticides, herbicides, fungicides, cyclamates, solvents, radiation and heat. For female animals, the list is shorter, but contains many of the chemicals grouped generically, as above.

The list for humans is shorter still and more tentative. Nevertheless, the stimulus to look further has come largely from two epidemiological studies of factory populations in the USA, which produced strong evidence that the nematocide, dibromochloropropane, and the chlorinated hydrocarbon insecticide, chlordecone, could cause human sterility.

It is important to remember that an abnormal pregnancy outcome spans a wide range of events in the timetable from ova/sperm to live birth.

4.4   Target organs

These include
- normal gonadal function
- union of ovum and sperm
- placental implantation
- embryonic organogenesis (1–3 months *in utero*)
- fetal development (3–9 months *in utero*)
- normal birth
- healthy childhood (no cancers).

Furthermore, at least 15–20% of normal conceptions fail to reach full growth and normal delivery. These could be termed 'normal' spontaneous abortions though most that have been analysed show gross genetic abnormalities. Congenital malformations have been reported in about 3% of all newborn children with a further 3% reported during postnatal or later development. Some childhood cancers are traceable to maternal exposure during pregnancy, e.g. vaginal carcinoma resulting from mothers treated during pregnancy with diethylstilboestrol. Most abnormal pregnancy outcomes, however, have no known cause.

Table 4.7 lists a number of chemicals (or occupations) where the link with abnormal pregnancies is real or suspected.

The evidence for suggesting that occupational exposure can disrupt other endocrine functions is largely lacking, except for one area—the

Table 4.7 Suspected reproductive effects of occupational exposures

Agent	Fertility	Spontaneous Abortion	Prematurity	Birth defects
Lead	√	?	—	?
Mercury	?	?	—	√ (Methyl)
Cadmium	√	—	—	—
Dibromochloropropane	√	—	—	—
Chlordecone	√	—	—	—
PCBs	—	—	√	√
Organic solvents	—	—	—	?
Oral contraceptives	√	—	—	—
Anaesthetic gases	—	?	—	—
Heat	√	—	—	—
Ethylene oxide	—	?	—	—
Cytotoxic drugs	√	√	—	—
Ionizing radiation	√	—	√	√
Heavy work	—	√	?	—
Shift work	—	?	—	—
Non-ionizing radiation	—	?	—	?
Laboratory work	—	?	—	?

theoretical possibility that any endocrinologically active pharmaceutical preparation, such as betamethasone, thyroxine and the anticancer drugs, could affect the workers who formulate or synthesize it. There is anecdotal evidence from the industry that such effects can occur but good epidemiological studies have not been executed.

Appendix 4.1   Prescribed diseases in the UK

Prescribed disease or injury	Occupation *Any occupation involving:*
A CONDITIONS DUE TO PHYSICAL AGENTS	
A1 Inflammation, ulceration or malignant disease of the skin or subcutaneous tissues or of the bones, or blood dyscrasia, or cataract, due to electromagnetic radiations (other than radiant heat), or to ionizing particles.	Exposure to electromagnetic radiations (other than radiant heat) or to ionizing particles.
A2 Heat cataract.	Frequent or prolonged exposure to rays from molten or red-hot material.
A3 Dysbarism, including decompression sickness, barotrauma and osteonecrosis.	Subjection to compressed or rarefied air or other respirable gases or gaseous mixtures.
A4 Cramp of the hand or forearm due to repetitive movements.	Prolonged periods of handwriting, typing or other repetitive movements of the fingers, hand or arm.
A5 Subcutaneous cellulitis of the hand (beat hand).	Manual labour causing severe or prolonged friction or pressure on the hand.
A6 Bursitis or subcutaneous cellulitis arising at or about the knee due to severe or prolonged external friction or pressure at or about the knee (beat knee).	Manual labour causing severe or prolonged external friction or pressure at or about the knee.
A7 Bursitis or subcutaneous cellulitis arising at or about the elbow due to severe or prolonged external friction or pressure at or about the elbow (beat elbow).	Manual labour causing severe or prolonged external friction or pressure at or about the elbow.

4   Occupational diseases

Appendix 4.1   Prescribed diseases in the UK

Prescribed disease or injury	Occupation *Any occupation involving:*
A8 Traumatic inflammation of the tendons of the hand or forearm, or of the associated tendon sheaths.	Manual labour, or frequent or repeated movements of the hand or wrist.
A9 Miner's nystagmus.	Work in or about a mine.
A10 Substantial sensorineural hearing loss amounting to at least 50 dB in each ear, being due in the case of at least one ear to occupational noise, and being the average of pure tone losses measured by audiometry over the 1, 2 and 3 kHz frequencies (occupational deafness).	(a) The use of powered (but not hand-powered) grinding tools on cast metal (other than weld metal) or on billets or blooms in the metal producing industry, or work wholly or mainly in the immediate vicinity of those tools whilst they are being so used; or (b) the use of pneumatic percussive tools on metal, or work wholly or mainly in the immediate vicinity of those tools whilst they are being so used; or (c) the use of pneumatic percussive tools for drilling rock in quarries or underground or in mining coal, or work wholly or mainly in the immediate vicinity of those tools whilst they are being so used; or (d) work wholly or mainly in the immediate vicinity of plant (excluding power press plant) engaged in the forging (including drop stamping) of metal by means of closed or open dies or drop hammers; or (e) work in textile manufacturing where the work is undertaken wholly or mainly in rooms or sheds in which there are machines engaged in—weaving man-made or natural (including mineral) fibres or in the high-speed false twisting of fibres; or (f) the use of, or work wholly or mainly in the immediate vicinity of, machines engaged in cutting, shaping or cleaning metal nails; or (g) the use of, or work wholly or mainly in the immediate vicinity of, plasma spray guns engaged in the deposition of metal; or (h) the use of, or work wholly or mainly in the immediate vicinity of, any of the

4 Occupational diseases

Appendix 4.1   Prescribed diseases in the UK

Prescribed disease or injury	Occupation *Any occupation involving:*
	following machines engaged in working of wood or material composed partly of wood, that is to say: multi-cutter moulding machines, planing machines, automatic or semi-automatic lathes, multiple cross-cut machines, automatic shaping machines, double-end tenoning machines, vertical spindle moulding machines (including high-speed routing machines), edge banding machines, bandsawing machines with a blade width of not less than 75 millimetres and circular sawing machines in the operation of which the blade is moved towards the material being cut; or (i) the use of chain saws in forestry.
A11 Episodic blanching, occurring throughout the year, affecting the middle or proximal phalanges or in the case of a thumb the proximal phalanx, of (a) in the case of a person with 5 fingers (including thumbs) on one hand, any 3 of those fingers, or (b) in the case of a person with only 4 such fingers, any 2 of those fingers, or (c) in the case of a person with less than 4 such fingers, any one of those fingers or, as the case may be, the one remaining finger (vibration white finger).	(a) The use of hand-held chain saws in forestry; or (b) the use of hand-held rotary tools in grinding or in the sanding or polishing of metal, or the holding of material being ground, or metal being sanded or polished, by rotary tools; or (c) the use of hand-held percussive metal-working tools, or the holding of metal being worked upon by percussive tools, in riveting, caulking, chipping, hammering, fetting or swaging; or (d) the use of hand-held powered percussive drills or hand-held powered percussive hammers in mining, quarrying, demolition, or on roads or footpaths, including road construction; or (e) the holding of material being worked upon by pounding machines in shoe manufacture.
B CONDITIONS DUE TO BIOLOGICAL AGENTS	
B1 Anthrax.	Contact with animals infected with anthrax or the handling (including the loading or unloading or transport) of animal products or residues.

4   Occupational diseases

Appendix 4.1   Prescribed diseases in the UK

Prescribed disease or injury	Occupation *Any occupation involving:*
B2 Glanders.	Contact with equine animals or their carcases.
B3 Infection by *Leptospira*.	(a) Work in places which are, or are liable to be, infested by rats, field mice or voles, or other small mammals; or (b) work at dog kennels or the care or handling of dogs; or (c) contact with bovine animals or their meat products or pigs or their meat products.
B4 Ankylostomiasis.	Work in or about a mine.
B5 Tuberculosis.	Contact with a source of tuberculous infection.
B6 Extrinsic allergic alveolitis (including farmer's lung).	Exposure to moulds or fungal spores or heterologous proteins by reason of employment in: (a) agriculture, horticulture, forestry, cultivation of edible fungi or malt-working; or (b) loading or unloading or handling in storage mouldy vegetable matter or edible fungi; or (c) caring for or handling birds; or (d) handling bagasse.
B7 Infection by organisms of the genus *Brucella*.	Contact with (a) animals infected by *Brucella,* or their carcases or parts thereof, or their untreated products; or (b) laboratory specimens or vaccines of, or containing, *Brucella*.
B8 Viral hepatitis.	Contact with (a) human blood or human blood products; or (b) a source of viral hepatitis.
B9 Infection by *Streptococcus suis*.	Contact with pigs infected by *Streptococcus suis*, or with the carcases, products or residues of pigs so infected.

Appendix 4.1   Prescribed diseases in the UK

Prescribed disease or injury	Occupation *Any occupation involving:*
B10 (a) Avian chlamydiosis.	Contact with birds infected with *Chlamydia psittaci,* or with the remains or untreated products of such birds.
(b) Ovine chlamydosis.	Contact with sheep infected with *Chlamydia psittaci,* or with the remains or untreated products of such sheep.
B11 Q fever.	Contact with animals, their remains or their untreated products.
B12 Orf.	Contact with sheep or goats, their remains or their untreated products.
B13 Hydatidosis.	Contact with dogs for shepherds, veterinary workers and others who work with dogs.
C CONDITIONS DUE TO CHEMICAL AGENTS	
C1 Poisoning by lead or a compound of lead.	The use or handling of, or exposure to the fumes, dust or vapour of, lead or a compound of lead, or a substance containing lead.
C2 Poisoning by manganese or a compound of manganese.	The use or handling of, or exposure to the fumes, dust or vapour of, manganese, or a substance containing manganese.
C3 Poisoning by phosphorus or an inorganic compound of phosphorus or poisoning due to the anti-cholinesterase or pseudo anti-cholinesterase action of organic phosphorus compounds.	The use or handling of, or exposure to the fumes, dust or vapour of, phosphorus or a compound of phosphorus, or a substance containing phosphorus.
C4 Poisoning by arsenic or a compound of arsenic.	The use or handling of, or exposure to the fumes, dust or vapour of, arsenic or a compound of arsenic, or a substance containing arsenic.
C5 Poisoning by mercury or a compound of mercury.	The use or handling of, or exposure to the fumes, dust or vapour of, mercury or a compound of mercury, or a substance containing mercury.

4 Occupational diseases

Appendix 4.1   Prescribed diseases in the UK

Prescribed disease or injury	Occupation *Any occupation involving:*
C6 Poisoning by carbon bisulphide.	The use or handling of, or exposure to the fumes or vapour of, carbon bisulphide or a compound of carbon bisulphide, or a substance containing carbon bisulphide.
C7 Poisoning by benzene or a homologue of benzene.	The use or handling of, or exposure to the fumes of, or vapour containing, benzene or any of its homologues.
C8 Poisoning by a nitro- or amino- or chloro-derivative of benzene or of a homologue of benzene, or poisoning by nitrochlorbenzene.	The use or handling of, or exposure to the fumes of, or vapour containing, a nitro- or amino- or chloro-derivative of benzene, or of a homologue of benzene, or nitrochlorbenzene.
C9 Poisoning by dinitrophenol or a homologue of dinitrophenol or by substitute dinitrophenols or by the salts of such substances.	The use or handling of, or exposure to the fumes of, or vapour containing, dinitrophenol or a homologue or substituted dinitrophenols or the salts of such substances.
C10 Poisoning by tetrachloroethane.	The use or handling of, or exposure to the fumes of, or vapour containing, tetrachloroethane.
C11 Poisoning by diethylene dioxide (dioxan).	The use or handling of, or exposure to the fumes of, or vapour containing, diethylene dioxide (dioxan).
C12 Poisoning by methyl bromide.	The use or handling of, or exposure to the fumes of, or vapour containing, methyl bromide.
C13 Poisoning by chlorinated naphthalene.	The use or handling of, or exposure to the fumes of, or dust or vapour containing, chlorinated naphthalene.
C14 Poisoning by nickel carbonyl.	Exposure to nickel carbonyl gas.
C15 Poisoning by oxides of nitrogen.	Exposure to oxides of nitrogen.
C16 Poisoning by *Gonioma kamassi* (African boxwood)	The manipulation of *Gonioma kamassi* or any process in or incidental to the manufacture of articles therefrom.

4   Occupational diseases

Appendix 4.1   Prescribed diseases in the UK

Prescribed disease or injury	Occupation *Any occupation involving:*
C17 Poisoning by beryllium or a compound of beryllium	The use or handling of, or exposure to the fumes, dust or vapour of, beryllium or a compound of beryllium, or a substance containing beryllium.
C18 Poisoning by cadmium.	Exposure to cadmium dust or fumes.
C19 Poisoning by acrylamide monomer.	The use or handling of, or exposure to, acrylamide monomer.
C20 Dystrophy of the cornea (including ulceration of the corneal surface) of the eye.	(a) The use or handling of, or exposure to arsenic, tar, pitch, bitumen, mineral oil (including paraffin), soot or any compound, product or residue of any of these substances, except quinone or hydroquinone; or (b) exposure to quinone or hydroquinone during their manufacture.
C21 (a) Localized new growth of the skin, papillomatous or keratotic; (b) squamous-celled carcinoma of the skin.	The use or handling of, or exposure to, arsenic, tar, pitch, bitumen, mineral oil (including paraffin), soot or any compound, product or residue of any of these substances, except quinone or hydroquinone.
C22 (a) Carcinoma of the mucous membrane of the nose or associated air sinuses; (b) primary carcinoma of a bronchus or of a lung.	Work in a factory where nickel is produced by decomposition of a gaseous nickel compound which necessitates working in or about a building or buildings where that process or any other industrial process ancillary or incidental thereto is carried on.
C23 Primary neoplasm (including papilloma, carcinoma-in-situ and invasive carcinoma) of the epithelial lining of the urinary tract (renal pelvis, ureter, bladder and urethra).	(a) Work in a building in which any of the following substances is produced for commercial purposes: (i) alpha-naphthylamine, beta-naphthylamine or methylene-bis-orthochloroaniline; (ii) diphenyl substituted by at least one nitro or primary amino group or by at least one nitro and primary amino group (including benzidine);

Appendix 4.1   Prescribed diseases in the UK

Prescribed disease or injury	Occupation *Any occupation involving:*
	(iii) any of the substances mentioned in sub-paragraph (ii) above if further ring substituted by halogeno, methyl or methoxy groups, but not by other groups; (iv) the salts of any of the substances mentioned in the sub-paragraphs (i) to (iii) above; (v) auramine or magenta; or (b) the use or handling of any of the substances mentioned in sub-paragraph (a)(i) to (iv), or work in a process in which any such substance is used, handled or liberated; or (c) the maintenance or cleaning of any plant or machinery used in any such process as is mentioned in sub-paragraph (b), or the cleaning of clothing used in any such building as is mentioned in sub-paragraph (a) if such clothing is cleaned within the works of which the building forms a part or in a laundry maintained and used solely in connection with such works.
C24 (a) Angiosarcoma of the liver; (b) osteolysis of the terminal phalanges of the fingers; (c) non-cirrhotic portal fibrosis.	(a) Work in or about machinery or apparatus used for the polymerization of vinyl chloride monomer, a process which, for the purposes of this provision, comprises all operations up to and including the drying of the slurry produced by the polymerization and the packaging of the dried product; or (b) work in a building or structure in which any part of that process takes place.
C25 Occupational vitiligo.	The use or handling of, or exposure to, *para*-tertiary-butylphenol, *para*-tertiary-butylcatechol, *para*-amyl-phenol, hydroquinone or the monobenzyl or monobutyl ether of hydroquinone.
C26 Damage to the liver or kidneys due to exposure to carbon tetrachloride.	The use or handling of, or exposure to the fumes of, or vapour containing, carbon tetrachloride.

4   Occupational diseases

Appendix 4.1   Prescribed diseases in the UK

Prescribed disease or injury	Occupation *Any occupation involving:*
C27 Damage to the liver or kidneys due to exposure to trichloromethane (chloroform).	The use or handling of, or exposure to the fumes of, or vapour containing, trichloromethane (chloroform).
C28 Central nervous system dysfunction and associated gastrointestinal disorders due to exposure to chloromethane (methyl chloride.)	The use or handling of, or exposure to the fumes of, or vapour containing, chloromethane (methyl chloride).
C29 Peripheral neuropathy due to exposure to *n*-hexane or methyl *n*-butyl ketone.	The use or handling of, or exposure to the fumes of, or vapour containing, *n*-hexane or methyl *n*-butyl ketone.
D MISCELLANEOUS CONDITIONS	
D1 Pneumoconiosis.	*Any occupation* (a) set out in Part II of this Schedule; (b) specified in regulation 2(b)(ii).
D2 Byssinosis.	*Any occupation involving:* Work in any room where any process up to and including the weaving process is performed in a factory in which the spinning or manipulation of raw or waste cotton or of flax, or the weaving of cotton or flax, is carried on.
D3 Diffuse mesothelioma (primary neoplasm of the mesothelium of the pleura or of the pericardium or of the peritoneum).	(a) The working or handling of asbestos or any admixture of asbestos; or (b) the manufacture or repair of asbestos textiles or other articles containing or composed of asbestos; or (c) the cleaning of any machinery or plant used in any of the foregoing operations and of any changers, fixtures and appliances for the collection of asbestos dust; or (d) substantial exposure to the dust arising from any of the foregoing operations.
D4 Inflammation or ulceration of the mucous membrane of the upper respiratory passages or mouth produced by dust, liquid or vapour.	Exposure to dust, liquid or vapour.

4 Occupational diseases

Appendix 4.1   Prescribed diseases in the UK

Prescribed disease or injury	Occupation *Any occupation involving:*
D5 Non-infective dermatitis of external origin (including chrome ulceration of the skin but excluding dermatitis due to ionizing particles or electromagnetic radiations other than radiant heat).	Exposure to dust, liquid or vapour or any other external agent capable of irritating the skin (including friction or heat but excluding ionizing particles or electromagnetic radiations other than radiant heat).
D6 Carcinoma of the nasal cavity or associated air sinuses (nasal carcinoma).	(a) Attendance for work in or about a building where wooden goods are manufactured or repaired; or (b) attendance for work in a building used for the manufacture of footwear or components of footwear made wholly or partly of leather or fibre board; or (c) attendance for work at a place used wholly or mainly for the repair of footwear made wholly or partly of leather or fibre board.
D7 Asthma which is due to exposure to any of the following agents: (a) isocyanates; (b) platinum salts; (c) fumes or dusts arising from the manufacture, transport or use of hardening agents (including epoxy resin curing agents) based on phthalic anhydride, tetrachlorophthalic anhydride, trimellitic anhydride or triethylenetetramine; (d) fumes arising from the use of rosin as a soldering flux; (e) proteolytic enzymes; (f) animals including insects and other arthropods used for the purposes of research or education or in laboratories; (g) dusts arising from the sowing, cultivation, harvesting, drying, handling, milling, transport or storage of barley, oats, rye, wheat or maize, or the handling, milling, transport or storage of meal or flour made therefrom; (h) antibiotics; (i) cimetidine; (j) wood dust; (k) ispaghula;	Exposure to any of the agents set out in column 1 of this paragraph.

4 Occupational diseases

Appendix 4.1   Prescribed diseases in the UK

Prescribed disease or injury	Occupation *Any occupation involving:*
(l) castor bean dust; (m) ipecacuanha; (n) azodicarbonamide (o) glutaraldehyde; (p) persulphate salts or henna arising from their use in the hairdressing trade; (q) crustaceans or fish or products arising from these in the food processing industry; (r) reactive dyes; (s) soya bean; (t) tea dust; (u) green coffee bean dust; (v) fumes from stainless steel welding; and (z) any other sensitizing agent inhaled at work (occupational asthma).	
D8 Primary carcinoma of the lung where there is accompanying evidence of one or both of the following: (a) asbestosis; (b) bilateral diffuse pleural thickening.	(a) The working or handling of asbestos or any admixture of asbestos; or (b) the manufacture or repair of asbestos textiles or other articles containing or composed of asbestos; or (c) the cleaning of any machinery or plant used in any of the foregoing operations and of any chambers, fixtures and appliances for the collection of asbestos dust; or (d) substantial exposure to the dust arising from any of the foregoing operations.
D9 Bilateral diffuse pleural thickening.	(a) The working or handling of asbestos or any admixture of asbestos; or (b) the manufacture or repair of asbestos textiles or other articles containing or composed of asbestos; or (c) the cleaning of any machinery or plant used in any of the foregoing operations and of any chambers, fixtures and appliances for the collection of asbestos dust; or (d) substantial exposure to the dust arising from any of the foregoing operations.

4   Occupational diseases

Appendix 4.1   Prescribed diseases in the UK

Prescribed disease or injury	Occupation *Any occupation involving:*
D10 Lung cancer.	(a) Work underground in a tin mine; or (b) exposure to bis(chloromethyl) ether produced during the manufacture of chloromethyl methyl ether; or (c) exposure to zinc chromate calcium chromate or strontium chromate in their pure forms.

5 Chemical agents

5.1 Introduction, 125

5.2 Inorganic chemicals, 127

5.3 Organic chemicals, 138

5.4 Toxic gases, 158

5.1   Introduction

This chapter is concerned with specific chemical agents that may be confronted in the workplace, although it is impossible in a book of this size to cover all known ones. For example, in the USA, nearly 60 000 materials are listed under the Toxic Substances Control Act and NIOSH names over 39 000 chemicals in its Registry of Toxic Effects of Chemical Substances. Thus, the authors have adopted an alternative approach and provided a selection of the more important toxic agents. This should provide the reader with a taste, so to speak, for the toxicology. The sources listed in the further reading at the end of the book can provide a more substantial account.

The chemical agents are classified into the following categories: inorganic chemicals, organic chemicals and toxic gases. Where appropriate, the workplace standards and any relevant statutory requirements, codes of practice or guidance notes, are quoted. Note that the Control of Substances Hazardous to Health Regulations 1988 apply to all substances listed. In addition, the Reporting of Diseases and Dangerous Occurrences Regulations 1985 cover virtually all the prescribed diseases (Section 4.2), the notable exceptions being dermatitis, noise-induced hearing loss and repetitive strain injuries. These legal requirements will not be included under each substance.

Details of standard-setting procedures are to be found in Section 3.5.

Workplace environmental standards

Some cautionary remarks are required. The reader is urged neither to treat the quoted standard as the strict dividing line between what is safe and what is not, nor to compare the relative toxicity of one substance with another by means of the standard. All standards should be treated with caution since, in setting them, difficulties arise in the following major areas:

1   Defining the health risk on which to base the standard.
2   Acquiring research data from which to draw a conclusion.
3   Defining the population that is being exposed, bearing in mind the susceptibility and vulnerability of individuals within a group which may be affected by genetic, social and ethnic backgrounds.
4   Understanding the economic and social implications that may be consequent upon defining a particular standard.
5   Obtaining a reliable method of measuring the degree of hazard.
6   Defining the duration of exposure and rate of work.

5 Chemical agents

5.1 Introduction

7 Considering the interaction of more than one hazard acting simultaneously.

The HSE's list of occupational exposure limits quotes airborne concentrations either volumetrically as parts per million (p.p.m.) or gravimetrically as milligrams of the substance per cubic metre of air (mg m^{-3}). For many gases and vapours both units are used, the relationship between them being the airborne specific density of the substance, that is, by multiplying the p.p.m. value by the specific density of the airborne substance the mg m^{-3} will be obtained. For example: the vapour of toluene is 3.76 times heavier than air, thus an OEL of 50 p.p.m. becomes 188 mg m^{-3}. For many substances two categories of standard are quoted:

1 A long-term time-weighted average (TWA), normally for an 8-hour period.

2 A short-term exposure limit (STEL), normally for a 10-minute period.

Those which are capable of being absorbed through the skin are indicated by the notation Sk.

To understand the complete definitions of these values, the relevant publication of the Health and Safety Executive and the ACGIH should be consulted. Also, to appreciate how much or how little information has been used in deciding upon a particular value, the publication *Documentation of the recommended TLVs*, published by ACGIH, should be consulted.

With regard to the standards quoted in this chapter, HSE occupational exposure limits (OESs, MELs) have been taken from EH 40/91. Where no HSE figure is available, the ACGIH recommended limit has been quoted.

Monitoring of the workplace environment
This is amply covered in other texts given in the further reading list. Basic techniques for measuring the airborne concentration of a specific chemical or particle are, therefore, only briefly mentioned in this chapter, under the particular substance. It must be pointed out that these are not necessarily the definitive techniques, as often more than one is available and new methods are continually being developed. Many analytical chemists also have their own preferences and they should always be consulted before embarking upon a particular method of sampling. The HSE publish guidance for the analysis of certain substances in the series Methods for the Determination of Hazardous Substances (MDHS). Where appropriate, the MDHS numbers are quoted.

5.2   Inorganic chemicals

Antimony (Sb)

Occurrence: Metalliferous ores usually as sulphide (Sb_2S_3).

Properties: Silvery-white, soft metal with properties very similar to arsenic. Stibine (SbH_3) is a gas (p. 165).

Uses: Alloys, paint pigment, rubber compounding.

Metabolism: Few severe poisonings—probably similar to arsenic.

Health effects: Similar to arsenic, though vomiting, eye and mucous membrane irritation may be more severe. Cardiac arrhythmias and mild jaundice have been reported.

Biological monitoring: Urinary antimony, electrocardiography.

Treatment: BAL (British anti-lewisite, dimercaprol), i.m.

Measurement: Antimony and compounds (particulate): sampled on cellulose acetate filter of pore size 0.8 μm, at an air-flow rate of around 2 l min^{-1} for subsequent analysis for Sb, using atomic adsorption spectrophotometry.

Control standards: Antimony and compounds as Sb, HSE, OES: 0.5 mg m^{-3}.

HSE guidance: EH19 Antimony—Health and Safety Precautions.

Arsenic (As)

Occurrence: Widely dispersed in nature, usually in association with metalliferous ore, e.g. FeAsS. It is, therefore, a by-product of both ferrous and non-ferrous smelting, mainly as the trioxide As_2O_3.

Properties: Arsenic is a steel-grey brittle metal, As_2O_3 a crystalline solid. Trivalent and pentavalent forms, Arsine (AsH_3) is a gas (p. 165).

Uses: Alloys, insecticides, fungicides, rodenticides, pigments, decolorizer in glass- and paper-making.

Metabolism: Normal body constituent due to wide dispersion in nature. Stored in keratin. Excretion slow in urine.

Health effects:

Acute: severe respiratory irritation, headache, abdominal pain, diarrhoea and vomiting → shock. Skin irritation and allergy are possible. See also arsine (p. 165).

Chronic: Gastrointestinal symptoms occasionally. Peripheral neuropathy—mainly sensory. Dermatitis, with or without areas of depigmentation. Liver damage (?)—vascular or parenchymal in type. Carcinogenic changes in skin and lungs.

Health surveillance and biological monitoring: Arsenic levels in urine, hair and nails less reliable. 'Normal' levels: urine, 70 nmol mmol^{-1} creatinine, hair 5 p.p.m. Lung function tests.

Treatment: Non-specific for skin and respiratory disturbances. British anti-lewisite (BAL), i.m.

Measurement: Arsenic and compounds (particulate): sampled on cellulose acetate filter of pore size 0.8 μm at an air-flow rate of around 2 l min^{-1}, for subsequent analysis for As using atomic adsorption spectrophotometry.

Control standards:

Arsenic and compounds (except arsine and lead arsinate) as As, HSE MEL: 0.1 mg m^{-3}.

Lead arsenate as Pb, HSE MEL: 0.15 mg m^{-3}.

Arsine (see p. 165).

HSE guidance:

Arsenic and compounds: EH8—Arsenic: Health and Safety Precautions. MDHS 41.

Beryllium (Be)

Occurrence: Mainly as beryllium aluminium silicate, $3BeO.Al_2O_3.6SiO_2$. Emerald (with chromium oxide), aquamarine and chrysoberyl are three varieties.

Properties: A very light, hard, non-corrosive, grey metal.

Uses: Alloys, nuclear reactors (fission moderator, neutron source, with uranium), aerospace, ceramics, formerly used in fluorescent light tubes.

Metabolism: Absorption is poor from the gut but good from the lungs. Protein-bound with liver, spleen and skeleton deposition. Urinary excretion variable.

Health effects:

Acute: chemical pneumonitis—cough, chest pain, dyspnoea, pneumonia. Conjuctivitis, rhinitis, pharyngitis. Skin irritant.

Chronic: Sarcoid-like granulomata—mainly in the lungs but occasionally subcutaneous. The lung lesions can lead to progressive interstitial fibrosis with hilar lymphadenopathy $\rightarrow$ cor pulmonale. Beryllium is a suspect lung carcinogen. Chest radiography in severe cases shows widespread nodules 1–5 mm in size, which may coalesce.

Biological monitoring: Chest radiography. Pulmonary function tests.

Treatment: Non-specific management of pulmonary fibrosis. Chelation therapy with EDTA has been used with some success in acute poisoning and corticosteroids may be helpful in the chronic fibrotic lung disease.

Measurement: Sampled on to cellulose acetate filter of pore size 0.8 μm, with an air-flow rate of around 1 l min^{-1}, for subsequent analysis for Be using atomic adsorption spectrophotometry (30 mins at 0.025 mg m^{-3}).

Control standards: HSE OES: 0.002 mg m^{-3} (under review).
HSE guidance:
EH13—Beryllium: Health and Safety Precautions.
MDHS 29.

Cadmium (Cd)
Occurrence: Cadmium sulphide (CdS), usually in association with zinc ore.
Properties: Soft, ductile, silver-white metal, which is corrosion-resistant
  and electropositive.
Uses: Alloys, electroplating, alkaline storage batteries, pigments, nuclear
  reactors (neutron absorber).
Metabolism: Mainly absorbed through inhalation. Bound to plasma globulin
  with accumulation in the kidney with lesser amounts in the liver. Urinary
  excretion poor in the absence of renal damage.
Health effects:
Acute: Increased salivation, nausea, vomiting $\rightarrow$ shock (ingestion).
  Cadmium fumes can cause a severe chemical penumonitis which can
  lead to pulmonary oedema and death. Mucous membrane irritation may
  also occur.
Chronic: Non-specific features include gastrointestinal disturbance, yellow
  rings on the teeth and anosmia. The main target organs are, however,
  the lungs and kidneys. Emphysema can be severe and is usually focal
  (due to reduced α-anti-tryspin?). Nephrotoxicity is usually manifested
  as tubular damage with proteinuria (especially β_2-microglobulins),
  glycosuria and aminoaciduria. Prostate and lung carcinoma as well
  as hypertension are implicated as sequelae of chronic cadmium
  exposure.
Health surveillance and biological monitoring: Lung function tests, urinalysis
  for protein and cadmium. Cadmium in urine is indicative of exposure but
  is a poor estimate of effect. Renal cortical cadmium
  levels are more reliable but difficult to measure (neutron activation
  analysis).
Treatment: Calcium EDTA is useful in acute poisoning. Chronic renal and
  pulmonary effects are often discovered at a late and irreversible stage.
Measurement: Sampled on to cellulose acetate filter of pore size 0.8 μm
  at an air-flow rate or 2 l min^{-1}, for subsequent analysis for Cd using
  atomic adsorption spectrophotometry.
Control standards: Cadmium dust and salts, cadmium oxide fume as Cd,
  HSE MEL: 0.05 mg m^{-3}.
HSE guidance: MDHS 10 and 11.

5   Chemical agents

5.2   Inorganic chemicals

Chromium (Cr)

Occurrence: Chromate ore ($FeO.Cr_2O_3$).

Properties: Hard, corrosion-resistant, grey metal. Di-, tri- and hexavalent states.

Uses: Alloys, electroplating, pigments.

Metabolism: Essential trace element. Better absorption for hexavalent than trivalent forms. Lung retention is considerable. Excretion mainly urinary.

Health effects: Hexavalent salts are irritant and corrosive, causing chronic skin ulcers and respiratory tract irritation. Chromate ore workers have an increased incidence of lung carcinoma, thought to be due to the slightly soluble hexavalent chromium compounds of strontium, calcium and zinc. Chromium platers exposed to chromic acid mist have also been reported to have an excess of lung cancer.

Biological monitoring: None of value.

Treatment: Remove chromium salts from skin. Non-specific treatment of pneumonitis.

Measurement:

Chromium, chromates, soluble chromic and chromous salts: sampled on to cellulose acetate filter of pore size 0.8 μm and at an air-flow rate of around 1.5 l min^{-1}, for subsequent analysis for Cr using atomic adsorption spectrophotometry.

Chromic acid and chromates: sampled on to PVC filter of pore size 5 μm and at an air-flow rate of around 1 l min^{-1}, for subsequent colorimetric analysis.

Control standards:

Chromium metal, chromium II and III compounds as Cr, HSE OES: 0.5 mg m^{-3}. Chromium VI compounds (except lead chromates) as Cr, HSE OES: 0.05 mg m^{-3} (under review).

Lead chromate as Pb, HSE MEL: 0.15 mg m^{-3}.

HSE guidance:

EH2—Chromium: Health and Safety Precautions.

EH6—Chromic Acid Concentrations in Air.

Chrome Ulceration: Epitheliomatous—Ulceration Order (SR and O, 1919, No. 1775).

MDHS 12 and 13.

Lead (Pb)

Occurrence: Mainly as the sulphide (PbS), in association with other metallic sulphates.

Properties: Soft, bluish-grey metal. Heavy, malleable, ductile. Inorganic and organic compounds.

130

5.2   Inorganic chemicals

Uses: Pipes, sheet metal, foil, ammunition, pigments, solders anti-knock additive to petrol (organic compound only).

Metabolism: Poorly absorbed through gut (10%) but dependent on calcium and iron content of diet. Pulmonary absorption more effective. Transported in form bound to red cell membrane and mainly stored in bone. Excretion mainly urinary.

Health effects:

Inorganic:

Acute: non-specific with lassitude, abdominal cramps and constipation, myalgia and anorexia.

Chronic: peripheral motor neuropathy (especially wrist drop) and anaemia are the main late manifestations, though disturbances of haem synthesis and a slowing of motor nerve conduction times can be detected soon after excessive absorption has commenced. Renal damage and encephalopathy are rare and usually confined to children.

Organic: Differs from inorganic in being primarily associated with psychiatric manifestations such as insomnia, hyperexcitability and even mania.

Biological monitoring (two kinds):

Absorption: blood-lead (should be less than 3 μmol l^{-1} in occupationally exposed workers).

Metabolic: urinary δ-amino laevulinic acid and red cell zinc (free erythrocyte) protoporphyrin concentration. (For organic lead absorption, urinary lead estimation is more useful.)

Treatment: If necessary, calcium EDTA or penicillamine can be given. The latter can be given orally and is, therefore, the treatment of choice. Organic lead poisoning does not respond to such chelation therapy.

Measurement:

Inorganic lead: Sampled on to cellulose acetate filter of pore size 0.8 μm at an air-flow rate of around 2 l min^{-1}, for subsequent analysis for Pb using atomic adsorption spectrophotometry. Note that filters should be partially covered, as with UKAEA-type holders.

Organic lead: Sampled through charcoal tube at an air-flow rate of 1000 ml min^{-1}, for subsequent analysis of Pb using atomic adsorption spectrophotometry.

Control standards:

Inorganic lead as Pb, HSE MEL: 0.15 mg m^{-3}.

Tetraethyl lead (as Pb), HSE MEL: 0.10 mg m^{-3}.

Legal requirements and HSE guidance:

Control of Lead at Work Regulations, 1980 (SI, 1980, No. 1248).

5   Chemical agents

5.2   Inorganic chemicals

Control of Lead at Work Code of Practice, revised June 1985.
EH28—Control of Lead: Air Sampling Techniques and Strategies.
EH29—Control of Lead: Outside Workers.
MDHS 6, 7, 8 and 18.

Manganese (Mn)

Occurrence: Widely occurring as MnO_2, $MnSiO_3$.

Properties: Reddish-grey, hard metal. Decomposes in water.

Uses: Alloys, dry-cell batteries, potassium permanganate, glass and
ceramics, matches.

Metabolism: Essential trace element. Poorly absorbed from gut, somewhat
better from lungs. Accumulates in kidney, liver and bone. Excretion
largely through gut. Transport in body is intracellular, with surprisingly
low cerebral concentrations.

Health effects:

Acute: Manganese oxide fume is a respiratory and mucous membrane
irritant.

Chronic: Slow onset (1–2 years) with headache, asthenia, poor sleep and
disturbed mental state. Neurological signs are primarily of the basal
ganglia $\rightarrow$ Parkinsonism.

Health surveillance and biological monitoring: Surveillance of CNS
functions, especially extrapyramidal system. No special tests.

Treatment: Calcium EDTA *before* permanent brain damage has occurred.
L-Dopa has been shown to be useful, at least in the short term.

Measurement: Sampled on to cellulose acetate filter of pore size 0.8 μm at
an air-flow rate of 1.5 l min^{-1}, for subsequent analysis for Mn using
atomic adsorption spectrophotometry.

Control standards:

Manganese and compounds (except fume, tetroxide and organic
manganese) as Mn, HSE OES: 5 mg m^{-3}.

Manganese fume as Mn, HSE OES: 1 mg m^{-3}.

Trimanganese tetraoxide, HSE OES: 1 mg m^{-3}.

Tricarbonyl (beta-cyclopentadienyl) manganese as Mn, OES: 0.1 mg m^{-3}.

Tricarbonyl (methylcyclopentadienyl) manganese as Mn, OES:
0.2 mg m^{-3}.

Mercury (Hg)

Occurrence: Mainly as sulphide ore (HgS), rarely as liquid metal.

Properties: Liquid at normal temperature and pressure. Therefore has a measurable vapour pressure. Mixes in unique fashion with other metals (amalgams).

Uses: Scientific instruments, amalgams, 'silvering', solders, pharmaceuticals, paints, seed dressings (organic compounds only), explosives.

Metabolism: Salts rapidly absorbed by all routes (metallic mercury poorly absorbed from gut). Inorganic salts more readily absorbed through gut and excreted by kidneys than organic compounds. Organics have predilection for the CNS.

Health effects:

Acute: Rare in industry. Febrile illness with pneumonitis. If severe, can cause oliguric renal failure.

Chronic: Slow onset with peculiar neuropsychiatric disorder (erethism) with features of anxiety neurosis, timidity and paranoia. Accompanied by gingivitis, excessive salivation, intention tremor, dermatographia, scanning speech. Upper motor neurone lesions and visual field constriction more commonly associated with organic mercurialism. Anterior capsule of the lens of the eye may be discoloured. Nephrotic syndrome.

Biological monitoring: Mercury in urine (preferably 24-hour specimen) or blood. Preferably distinguishing between 'total' and inorganic mercury. Upper level of acceptable excretion in medium to long-term exposure is 120 nmol mmol^{-1} creatinine. Electromyography.

Treatment: BAL, calcium EDTA—both more effective for inorganic mercurialism.

Measurement:

Inorganic compounds: sampled on to cellulose acetate filter of pore size 5.0 μm at an air-flow rate of around 2 l min^{-1}, for subsequent analysis using atomic adsorption spectrophotometry; note that filters should be partially covered, as with UKAEA holders.

Mercury vapour: measured with a direct reading instrument using ultraviolet light (interfered with by the presence of oil mist of vapour).

Organic compounds: sampled through adsorbent tube (chromosorb) at an air-flow rate of 50 ml min^{-1} for subsequent analysis for Hg using atomic adsorption spectrophotometry.

Control standards:

Mercury and compounds (except organic alkyls) as Hg, HSE OES: 0.05 mg m^{-3}.

Mercury alkyls as Hg, HSE OES: 0.01 mg m^{-3} Sk.

HSE guidance:

5   Chemical agents

5.2   Inorganic chemicals

EH17—Mercury: Health and Safety Precautions.
MS12—Mercury: Medical Surveillance.
MDHS 16 and 58.

Nickel (Ni)

Occurrence: Sulphide ore extracted by separation or Mond process (unique reaction of nickel with carbon monoxide to produce nickel carbonyl $Ni(CO)_4$) (see p. 161).
Properties: Hard, ductile, magnetic, silver-white metal. Low corrosion.
Uses: Alloys (especially with steel), electroplating, oil catalyst, coins, ceramics, batteries.
Metabolism: Poor absorption with wide bodily distribution, especially brain and lungs. Rapid excretion in urine and faeces.
Health effects:
Acute: Allergic contact dermatitis. Fume can cause pneuonitis.
Chronic: Carcinoma of the nose and nasal sinuses associated with exposure to nickel, though exact aetiological agent is unknown—nickel oxides and sulphides (?).
Health surveillance and biological monitoring: Pre-employment allergy screening.
Treatment: Non-specific.
Measurement: Sampled on to cellulose acetate filter of pore size 0.8 μm at an air-flow rate of around 2 l min^{-1}, for subsequent analysis for Ni using atomic adsorption spectrophotometry.
Control standards:
Nickel compounds as Ni
    soluble compounds, HSE OES: 0.1 mg m^{-3} (under review).
    insoluble compounds, HSE OES: 0.5 mg m^{-3} (under review).
Nickel carbonyl as Ni, HSE OES, STEL: 0.24 mg m^{-3}.
Nickel organic compounds as Ni, HSE OES: 1 mg m^{-3}.
MDHS 42.

Phosphorus (P)

Occurrence: Wide, usually as phosphates of calcium.
Properties: Three allotropic forms—yellow (spontaneously ignites), red and black. Can form a gaseous hydride (PH_3) (p. 165), as well as organic compounds (p. 150).
Uses: Agriculture, baking powder, detergents, explosives, paper and printing. Phosphoric acid is an anti-rust agent.
Metabolism: Rapid absorption by ingestion or inhalation.

5.2   Inorganic chemicals

Health effects:

Acute: Phosphorus oxides cause severe pneumonitis. Yellow phosphorus
    can cause severe burns and liver damage.

Chronic: 'Phossy jaw'—now virtually unknown. A severe, painful necrotic
    disease of the bone—usually the mandible.

Health surveillance: Dental surveillance, including radiography for early
    phossy jaw.

Treatment: Wound debridement.

Measurement:

Phosphoric acid: sampled on to cellulose acetate filter of pore size 0.8 μm
    at an air-flow rate of 1.5 l min^{-1}, for subsequent colorimetric analysis.

Phosphorus compounds: bubbled into distilled water for subsequent
    colorimetric analysis.

Control standards:

Phosphoric acid, ACGIH TLV: 1 mg m^{-3}.

Phosphorus (yellow), HSE OES: 0.1 mg m^{-3}.

Phosphorus oxychloride, ACGIH TLV: 0.6 mg m^{-3}.

Phosphorus pentachloride, HSE OES: 1 mg m^{-3}.

Phosphorus pentasulphide, HSE OES: 1 mg m^{-3}.

Phosphorus trichloride, HSE OES: 1.5 mg m^{-3}.

Phosphoryl trichloride, HSE OES: 1.2 mg m^{-3}.

Platinum (Pt)

Occurrence: Alluvial deposits.

Properties: Soft, ductile, malleable, non-corrosive, white metal.

Uses: Electrical contacts, catalyst, alloys, jewellery, photography.

Health effects:

Acute: nasal irritation.

Chronic: Platinum asthma (especially after exposure to chloroplatinic acid
    or one of its salts). Dry, scaly skin—irritant or allergic dermatitis.

Health surveillance: Pulmonary function tests. Some advocate pre-
    employment allergy testing.

Treatment: Non-specific.

Measurement: Sampled on to cellulose acetate filter of pore size 0.8 μm
    at an air-flow rate of around 2 l min^{-1}, for subsequent analysis using
    atomic adsorption spectrophotometry.

Control standards:

Platinum metal, HSE OES: 5 mg m^{-3}.

Platinum salts, soluble as Pt, HSE OES: 0.002 mg m^{-3} Sen.

MDHS 46.

5.2   Inorganic chemicals

Thallium (Tl)

Occurrence: As a complex with copper, silver and selenium, $(TlCuAg)_2Se$.

Properties: Soft, malleable, silver-grey metal. Soluble in acids. Oxidizes in air.

Uses: Rodenticide, insecticide, optical equipment, alloys, fireworks, dyes, depilatory agent.

Metabolism: Readily absorbed by all routes. Excretion is slow, poisoning is cumulative.

Health effects:

Acute: Vomiting, diarrhoea, abdominal pain, anxiety state. Acute ascending polyneuritis.

Chronic: Polyneuritis, alopecia, albuminuria, ocular lesions.

Health surveillance: Medical surveillance of peripheral and central nervous system. Hair loss is a significant sign.

Treatment: None.

Measurement: Sampled on to cellulose acetate filters of pore size 0.8 μm at an air-flow rate of around 2 l min^{-1}, for subsequent analysis using atomic adsorption spectrophotometry.

Control standards: Thallium, soluble compounds as Tl, HSE OES: 0.1 mg m^{-3}.

Vanadium (V)

Occurrence: Vanadinate ($9PbO.3V_2O_5PbCl_2$). Also a by-product of oil-burning furnaces, when vanadium pentoxide is deposited in the flues.

Properties: Grey-white lustrous powder.

Uses: Alloys with steel increase the hardness and malleability of products. Catalyst, insecticide, dyes.

Metabolism: Inhalation is main route of entry. Rapid renal excretion.

Health effects:

Acute: Severe pneumonitis (usually due to exposure to flue dust) with mucous membrane irritation and gastrointestinal disturbances.

Chronic: Chronic bronchitis. Eczematous skin lesions. Fine tremor of extremities. Black tongue.

Biological monitoring: Urinary vanadium.

Treatment: Non-specific.

Measurement: Sampled on to cellulose acetate filter of pore size 0.8 μm at an air-flow rate of around 2 l min^{-1}, for subsequent analysis using atomic adsorption spectrophotometry.

Control standards: Divanadium pentoxide as V, HSE OES: total inhalable dust 0.5 mg m^{-3}, fume and respirable dust 0.05 mg m^{-3}.

5   Chemical agents

5.2   Inorganic chemicals

Welding fumes

Welding fumes cannot be easily classified as the composition is dependent upon the alloy being welded and the method of welding being used and is invariably a mixture of substances. Arc welding fumes in particular must be analysed for individual constituents, both particulate and gaseous, to determine whether specific recommended limits are exceeded. Under the heading of 'Mixed exposures' in EH40/91 (para 45) the HSE refers the reader to Guidance Note EH54 *Assessment of exposure to fume from welding and allied processes.* In the ACGIH limit recommendation book a useful paragraph on the complex nature of welding fume is appended.

HSE guidance: MS 15 welding EH54.

Zinc (Zn)

Occurrence: As sulphide or carbonate.

Properties: High corrosion resistance. Poor conductor.

Uses: Galvanizing, brass (5–40% Zn), dyes, electroplating.

Metabolism: Essential trace element, e.g. carbonic anhydrase is a zinc-containing enzyme. Also thought to be an important factor in wound healing. Poorly absorbed. Faecal excretion.

Health effects:

Acute: Metal fume fever. Can occur with fumes of other metals but zinc is the most common. Symptoms resemble influenza and come on within 12 hours of exposure. Recovery is rapid with no sequelae. Zinc chloride is a skin and lung corrosive.

Health surveillance: Allergy surveillance. Pulmonary function tests.

Treatment: Non-specific management of pulmonary disorders. Acute skin contact with zinc chloride benefits from irrigation with calcium EDTA solution.

Measurement:

Zinc fume: sampled on to a cellulose acetate filters of pore size 0.8 μm at an air-flow rate of around 2 l min^{-1}, for subsequent analysis using atomic adsorption spectrophotometry.

Zinc stearate: sampled on to a weighed glass-fibre filter, for gravimetric analaysis.

Control standards:

Zinc chloride fume, HSE OES: 1 mg m^{-3}.

Zinc oxide fume, HSE OES: 5 mg m^{-3}.

Zinc distearate, HSE OES: total inhalable dust 10 mg m^{-3}, respirable dust 5 mg m^{-3}.

___

5.3   Organic chemicals

In general, organic chemicals are carbon-containing compounds. Carbon has a valency of 4 and a unique ability to form chain or ring structures with itself.

Aliphatic compounds are chains: $-C-C-C-$

e.g. methane

Aromatic compounds are rings:

e.g. benzene

or

(abbreviated)

The two groups have differing properties, depending not only on their configuration but also on the elements attached to the 'unoccupied' valency arms.

Industrially important organic compounds are frequently hydrocarbons (methane and benzene) or hydrocarbons with some hydrogen atoms replaced by halogens such as chlorine. The chlorinated hydrocarbons are fat-soluble and are often non-flammable, non-combustible and non-explosive but *not* safe.

As a *general* rule, increasing the chlorination of aliphatic hydrocarbons leads to an increasing toxicity, while the reverse is true of aromatic hydrocarbons. Many chlorinated hydrocarbons are hepatotoxic and, on combustion at high temperatures, release the toxic irritant gas, phosgene ($COCl_2$).

138

5 Chemical agents

5.3 Organic chemicals

Acrylamide

$$H_2C = CH$$

C with O and NH_2

Properties: White crystalline powder. Readily undergoes polymerization, water-soluble. MW 71.08.

Uses: Manufacture of flocculators, dyes, leather substitutes, paper, pigments; used in soil stabilization, mining and removal of industrial wastes.

Metabolism: Absorbed by inhalation, ingestion or (mainly) through the skin. Metabolism largely unknown. Monomer is neurotoxic, polymer is harmless.

Health effects:

Acute: Eye and mucous membrane irritation.

Chronic: Peripheral neuropathy and mid-brain lesions. Numbness, paraesthesia and weakness of limbs (legs more than arms). Ataxia, slurred speech, lethargy, increased sweating.

Health surveillance: Electromyography and nerve conduction studies.

Treatment: Wash contaminated skin thoroughly. No specific treatment for neurotoxic effects but further exposure may produce a more severe reaction dose for dose.

Measurement: Sampled on to a charcoal tube at an air-flow rate of 200 ml min^{-1}, for subsequent gas chromatographic analysis.

Control standards: HSE MEL: 0.3 mg m^{-3}Sk.

MDHS 57.

Acrylonitrile

$$H_2C = CH$$

C $\equiv$ N

Properties: Explosive, inflammable liquid. Readily polymerized. MW 53.06.

Uses: Manufacture of synthetic rubber, acrylic resins and fibres.

Metabolism: Skin and lung absorption. Toxicity due to the release of cyanide radicle (CN^{-}). Excreted as thiocyanate in urine.

5.3   Organic chemicals

Health effects:
Acute: Vapour is severe eye irritant and skin vesicant. Headache, sneezing, weakness, dizziness → asphyxia and death.

Chronic: Epidemiological studies in humans support the animal experiments that suggest that acrylonitrile is carcinogenic (probably lung, colon (?)).

Biological monitoring: Cyanmethaemoglobin levels, blood pH and bicarbonate.

Treatment: Must be rapid as for any cyanide poisoning.
Amylnitrite (inhalation). Dicobalt edetate (i.v.).
Hydroxycobalamine (?) (i.v.).

Measurement: Sampled on to a charcoal tube at an air-flow rate of 200 ml min^{-1}, for subsequent gas chromatographic analysis. Colorimetric detector tubes are also available.

Control standards: HSE MEL: 4 mg m^{-3} (2 p.p.m.) Sk.

HSE guidance: EH27—Acrylonitrile: Personal Protective Equipment. MDHS 1, 2 and 55.

Aniline

Properties: Colourless, oily liquid with aromatic odour. MW 93.13.

Uses: Dyes, perfumes, explosives, pharmaceuticals, rubber processing.

Metabolism: Skin and lung absorption. Converts haemoglobin to methaemoglobin with a resulting diminution of the oxygen-carrying capacity of blood.

Health effects:
Acute: Mild skin irritant. Moderate exposure may only cause some cyanosis. Severe poisoning results in anoxia and death, which may be delayed for a few hours after exposure.

Biological monitoring: Methaemoglobin levels must be monitored during treatment.

Treatment: Remove all traces of aniline from the skin and all contaminated clothing. Methylene blue (i.v.) is justified for comatose patients with methaemoglobin levels above 60%. Methylene blue reduces methaemoglobin to haemoglobin.

Measurement: Sampled on to a silica gel tube at an air-flow rate of

---

200 ml min^{-1}, for subsequent analysis using gas chromatography.
Colorimetric detector tubes are also available.
Control standards: HSE OES: 10 mg m^{-3}(2 p.p.m.) Sk (under review).

Benzene

(C_6H_6)

Occurrence: By-product of petroleum and coke-oven industries.
Properties: Colourless, inflammable liquid. The fat solvent *par excellence.*
  MW 78.11.
Uses: The initial compound used in the production of numerous organic
  aromatics, including styrene, phenol, cyclohexane, as well as many
  plastics, paints, glues, dyes and pharmaceuticals.
Metabolism: Lung and skin absorption with ready transportation and
  uptake by fatty tissue. Excretion is slow through the lungs, with a little
  appearing in the urine as conjugated phenols.
Health effects:
Acute: Uncommon in industry. Dizziness, light-headedness, vomiting $\rightarrow$
  unconsciousness and death.
Chronic: Bone marrow depression with a delayed effect, in some cases, of
  many years. The early symptoms and signs are vague but later
  tiredness and spontaneous bleeding may occur as anaemia,
  pancytopenia and/or thrombocytopenia become more severe. Aplastic
  anaemia, acute myeloblastic leukaemia and acute erythroleukaemia are
  the most feared effects of chronic exposure.
Health surveillance: Periodic red and white cell counts $\pm$ bone marrow
  analysis where indicated. Urinary phenol has been used for biological
  monitoring.
Treatment: Acute intoxication will usually respond to removal from
  exposure and respiratory system support measures. The development
  of aplastic or leukaemic effects augurs ill. Treatment is as for other
  causes of these conditions, but the prognosis is poor.
Measurement: Sampled on to charcoal tube at an air-flow rate of
  200 ml min^{-1}, for subsequent analysis using gas chromatography.
  Colorimetric detector tubes are also available.
Control standards: HSE MEL: 15 mg m^{-3} (5 p.p.m.).
MDHS 17 and 50.

5 Chemical agents

5.3 Organic chemicals

Carbon disulphide

Properties: Colourless liquid. MW 76.16.
Uses: Solvent for fats, sulphur, rubber oils. Insecticide.
Preparation of viscose rayon.
Metabolism: Absorbed through lungs and skin. Slow metabolism and
excretion with main concentration build-up in the brain.
Health effects:
Acute: Severe skin and mucous membrane irritant. Dizziness, headaches,
psychosis, drowsiness.
Chronic: Four distinct syndromes. (i) Parkinsonian-like affection due to
damage to the corpus striatum and globus pallidus. (ii) Peripheral
neuropathy affecting motor and sensory nerves as well as ocular
nerves. (iii) Psychotic conditions (rarely seen nowadays but lesser
neuropsychiatric states are still described). (iv) Cardiovascular disease—
possibly due to increased blood cholesterol and β-lipoprotein leading to
ischaemic heart disease and peripheral vascular damage. However,
recent research suggests that part (at least) of this cardiovascular effect
may be due to an acute myotoxic effect on cardiac muscle leading to
fatal arrhythmias.
Biological monitoring: The iodine-azide reaction with urine will detect
organic sulphate metabolites of carbon disulphide but the reaction is not
specific for CS_2.
Treatment: Symptomatic treatment only.
Measurement: Sampled on to a charcoal tube at an air-flow rate of
200 ml min^{-1}, for subsequent analysis using gas chromatography.
Colorimetric detection tubes are also available.
Control standards: HSE MEL: 30 mg m^{-3} (10 p.p.m.) Sk.
HSE guidance: MDHS 15, EH45.

Carbon tetrachloride

$$Cl - \underset{\underset{Cl}{|}}{\overset{\overset{Cl}{|}}{C}} - Cl$$

Properties: Colourless, non-inflammable liquid with characteristic smell.
Burning yields phosgene and hydrogen chloride gas. MW 153.82.

5.3   Organic chemicals

Uses: Solvent, degreaser, manufacture of refrigerants such as Freon, fire
   extinguisher and grain fumigant.

Metabolism: Absorbed through lungs, skin and gut and stored in fatty
   tissues. Excreted unchanged through the lungs, though some is
   metabolized and excreted in the urine.

Health effects:

Acute: Nausea, vomiting, drowsiness, dizziness.

Chronic: Dry, scaly dermatitis. Centrilobular necrosis $\pm$ fatty degeneration
   of the liver. Acute oliguric renal failure. There is a syngergistic effect if
   exposure is associated with alcohol intake.

Biological monitoring: Limited value of blood concentrations.

Treatment: Non-specific. Most cases recover but renal and hepatic damage
   may be permanent.

Measurement: Sampled on to a charcoal tube at an air-flow rate of
   1000 ml min^{-1}, for subsequent analysis using gas chromatography.
   Colorimetric detection tubes are also available.

Control standards: HSE OES: 65 mg m^{-3} (10 p.p.m.) Sk (under review).

HSE guidance: MDHS 28.

Chlorinated naphthalenes

$$C_{10}H_{(8-n)}Cl_n$$

Properties: A group of compounds with varying degrees of chlorination.
   The higher the chlorine content, the higher the melting point.

Uses: Wire insulation and flame resistance in condensers.

Metabolism: Inhalation of fumes and percutaneous absorption of liquids.

Health effects:

Acute: Little effect.

Chronic: Two distinct effects. (i) Chloracne from skin contact. (ii) Liver
   damage from inhalation.

Health surveillance: Liver function and skin examination.

Treatment: Non-specific.

Chloroform

$$Cl - \overset{\displaystyle H}{\underset{\displaystyle Cl}{C}} - Cl$$

Properties: Clear, colourless, non-flammable liquid with characteristic
   odour. MW 119.38.

5.3   Organic chemicals

Uses: Fat solvent, manufacture of fluorocarbons, plastics. Abandoned as an anaesthetic agent due to its hepatotoxicity.

Metabolism: Absorbed through the lungs (and skin). Stored in fatty tissue and slowly excreted through the lungs and, to a lesser extent, the kidneys.

Health effects:

Acute: Skin irritant. Potent anaesthetic.

Chronic: Liver enlargement and damage potentiated by alcohol abuse (causes hepatic tumours in rodents). Oliguric renal failure. Chronic dry, scaly dermatitis.

Biological monitoring: Blood chloroform level. Periodic liver function tests.

Treatment: Non-specific.

Measurement: Sampled on to charcoal tube at an air-flow rate of 1000 ml min^{-1}, for subsequent analysis using gas chromatography.

Control standard: HSE OES: 50 mg m^{-3} (10 p.p.m.) (under review). MDHS 28.

DDT (I,I,I-trichlorobis (chlorophenyl) ethane)

2 isomers

Properties: White crystalline solid. MW 354.49.

Uses: Insecticide.

Metabolism: Absorbed through gut and intact skin. Metabolized slowly to DDE and DDA with storage in fatty tissues and excretion in urine.

Health effects: In high doses (10 mg kg^{-1} body weight), DDT causes CNS effects including paraesthesia, tremors and convulsions. Occupational exposure studies have been remarkable for their lack of evidence of long-term CNS or hepatic effects, despite animal evidence of hepatic tumours and the widespread banning of the use of the compound.

Biological monitoring: DDA in urine.

Treatment: Clear skin contamination. Diazepam for convulsions. Otherwise, non-specific.

Measurement: Airborne particulate is sampled on to a glass fibre filter at an air-flow rate of around 1.5 l m^{-1} and is subsequently removed by iso-octane and the aliquot analysed by gas chromatography.

Control standard: HSE OES: 1 mg m^{-3}.

5   Chemical agents

5.3   Organic chemicals

Dinitrobenzene

ortho-   meta-   para-

Properties: Solid, colourless, odourless flakes. Three isomeric forms.
MW 168.11.
Uses: Dyes, explosives.
Metabolism: Inhalation and skin absorption causes methaemoglobinaemia.
Health effects:
Acute: Headache, dizziness, vomiting and weakness. Hypotension,
tachycardia, hyperpnoea.
Chronic: Weakness, fatigue, cyanosis and pallor. Symptoms may be
exacerbated by alcohol or sunlight. Liver damage (acute necrosis) is rare.
Health surveillance: Methaemoglobin levels. Occasionally patients exhibit
albuminuria or porphyrinuria.
Treatment: Remove all traces of DNB from the skin. Methylene blue may
be justified in severe methaemoglobinaemia; otherwise, oxygen and
other respiratory supportive measures usually suffice.
Measurement: Air is drawn through a cellulose acetate filter in series with a
bubbler containing 10 ml of ethylene glycol at a flow rate of $1.5 \, l \, m^{-1}$.
At the end of the sampling period, the filter is added to the bubbler
liquid. The resulting sample is analysed by high-pressure liquid
chromatography.
Control standard: HSE OES: $1 \, mg \, m^{-3}$ (0.15 p.p.m.).

Dinitrophenol

(2 isomers)

Properties: Explosive, yellow crystalline solid. MW 184.11.
Uses: Explosive, dyes, timber preservative.

5.3   Organic chemicals

Metabolism: Absorbed through the gastrointestinal tract, skin and
respiratory tract. Effects enhanced by heat and alcohol. Excretion is
slow, urine becomes orange and skin is turned yellow. DNP and its
homologue dinitro-orthocresol interfere with temperature regulation
with uncoupling of oxidative phosphorylation.

Health effects:

Acute: Sudden onset of chest pain and dyspnoea ± hyperpyrexia, profuse
sweating and thirst.

Chronic: Similar to acute effects ± liver tenderness and jaundice. Cataract
formation. Neutropenia and albuminuria were noted when DNP was in
vogue for weight reduction therapy.

Biological monitoring: Urine analysis for DNP or its metabolite 2-amino-4-
nitrophenol has limited use.

Treatment: Cooling, oxygen, sedation.

Formaldehyde

$$H-\overset{\displaystyle \overset{O}{\|}}{C}-H$$

Properties: Colourless gas with a pungent odour. Commonly used as an
aqueous solution (formalin) of 34–38% formaldehyde. MW 30.03.

Uses: Plastics and resin manufacture, preservative, intermediate in chemical
manufacture. Used also in textile industry as a crease-resistant agent.

Metabolism: Mainly by inhalation. Metabolized in liver and excreted in urine
and exhaled air. Converted to formate in many tissues, including red
blood cells.

Health effects:

Acute: Severe mucous membrane irritation.

Chronic: Potent allergen for skin, (?) respiratory tract. Probable animal
carcinogen (nasal tumours). No clear evidence to date of human
carcinogenic effect but formaldehyde and hydrochloric acid can produce
bischloromethylether (BCME), a proven human lung carcinogen.

Biological monitoring: None in common use.

Treatment: Symptomatic.

Measurement: Air is bubbled through a 0.5% solution of 3-methyl-2-
benzothiazolone hydrazone in a bubbler at a flow rate of $1\,l\,m^{-1}$. The
resulting solution is analysed by colorimetric means. Colorimetric
detector tubes are also available.

Control standards: HSE MEL: $2.5\,mg\,m^{-3}$ (2 p.p.m.).

MDHS 19.

5 Chemical agents

5.3 Organic chemicals

Isocyanates

(Toluene Di-isocyanate 2 isomers)

Properties: Colourless liquid. MW 174.16.

Uses: Polyurethane production varying from flexible form (toluene di-isocyanate: TDI) to rigid types (diphenylmethane di-isocyanate: MDI).

Metabolism: Poorly understood. Effects proportional to volatility (TDI > MDI).

Health effects:

Acute: Respiratory irritation in all exposed at high concentrations. Hypersensitivity with asthma $\pm$ skin disorders in the allergic minority.

Chronic: Permanent pulmonary disability and neuropsychiatric effects are postulated but disputed.

Health surveillance: Prick test $\pm$ bronchial provocation testing. Serial spirometric readings.

Treatment: Symptomatic. Avoid further exposure in sensitized individuals.

Measurement: Air is drawn through an impregnated tape. The resulting stain is photometrically examined. Proprietary instrument available. A colorimetric detection tube for TDI is available. Alternatively, a measured volume of sample air is drawn through a glass impinger (bubbler) containing 1-(2-methoxyphenyl)-piperazine for subsequent analysis using high-performance liquid chromatography (MDHS 25). In another method, a measured volume of sample air is drawn through a glass impinger containing dimethylformamide and dilute hydrochloric acid for subsequent colorimetric analysis (MDHS 49).

Control standards: HSE MEL for all isocyanates in air (as –NCO): 0.02 mg m^{-3}.

HSE guidance: EH16—Isocyanates: Toxic Hazards and Precautions.

MS8—Isocyanates: Medical Surveillance.

MDHS 25 and 49.

Ketones and ethers

Uses: Solvents (especially dimethylketone (acetone), methylethyl ketone (MEK), methylbutyl ketone (MBK), di-ethyl ether).

5   Chemical agents

5.3   Organic chemicals

Health effects:
Acute: Upper respiratory tract irritants. Cause dermatitis.
Chronic: Narcotic. MBK can cause peripheral neuropathy. Bischloro-
methylether (BCME) is a strong alkylating agent and a potent lung
carcinogen.
Biological monitoring: Non-specific.
Treatment: Non-specific.
Measurement: Ethers and ketones are sampled by drawing air through a
charcoal tube at a flow rate of 200 ml m^{-1}, for subsequent analysis
using gas chromatography.
Control standards: Allyl glycidylether (AGE), HSE OES: 22 mg m^{-3}
(5 p.p.m.) Sk.
n-Butyl glycidylether (BGE), HSE OES: 135 mg m^{-3} (25 p.p.m.).
Bis(chloromethyl)ether, HSE (BCME) OES: 0.001 p.p.m. (under review).
Diethylether, HSE OES: 1200 mg m^{-3} (400 p.p.m.).
Diglycidylether (DGE), HSE OES: 0.6 mg m^{-3} (0.1 p.p.m.).
Di-isopropylether, HSE OES: 1050 mg m^{-3} (250 p.p.m.).
Ethylene glycol monobutylether (2-butoxyethanol), HSE MEL:
120 mg m^{-3} (25 p.p.m.) Sk.
Ethylene glycol monomethylether (2-methoxyethanol), HSE MEL:
16 mg m^{-3} (5 p.p.m.) Sk.
Glycol monoethylether (2-ethoxyethanol), HSE MEL: 37 mg m^{-3}
(10 p.p.m.) Sk.
Isopropyl glycidylether, (IGE) HSE OES: 240 mg m^{-3} (50 p.p.m.).
Propylene glycoldinitrate (PGDN), HSE OES: 1.2 mg m^{-3} (0.2 p.p.m.) Sk.
Propylene glycolmonomethylether, HSE OES: 360 mg m^{-3} (100 p.p.m.) Sk.
Diethylketone (pentan-3-one), HSE OES: 700 mg m^{-3} (200 p.p.m.).
2,6-Dimethylheptan-4-one, HSE OES: 150 mg m^{-3} (25 p.p.m.).
Ethylamylketone (5-methylheptan-3-one), HSE OES: 130 mg m^{-3}
(25 p.p.m.).
Ethylbutylketone (heptan-3-one), HSE OES: 230 mg m^{-3} (50 p.p.m.).
Methyl-n-amylketone (heptan-2-one), HSE OES: 240 mg m^{-3} (50 p.p.m.).
Methylbutylketone (hexan-2-one) (MBK), HSE OES: 20 mg m^{-3} (5 p.p.m.) Sk.
Methylethylketone (butan-2-one) (MEK), HSE OES: 590 mg m^{-3}
(200 p.p.m.).
2-Methylcyclohexanone, HSE OES: 230 mg m^{-3} (50 p.p.m.) Sk.
Methylisoamylketone (5-methylhexan-2-one), HSE OES: 240 mg m^{-3}
(50 p.p.m.).
Methylisobutylketone (methylpentan-2-one) (MIBK), HSE OES:
205 mg m^{-3} (50 p.p.m.) Sk.

5.3   Organic chemicals

Methylisopropylketone (MIPK), ACGIH rec. limit: 705 mg m^{-3}
    (200 p.p.m.).
Methylpropylketone (pentan-2-one), HSE OES: 700 mg m^{-3} (200 p.p.m.).
HSE guidance: MDHS 23.

Methyl alcohol (methanol)

$$H-\overset{\overset{\displaystyle H}{|}}{\underset{\underset{\displaystyle H}{|}}{C}}-OH$$

Properties: Colourless liquid which smells like ethanol. MW 32.04.
Uses: Celluloid manufacture, paint remover, varnishes, antifreeze, cements.
Metabolism: Absorbed by all routes and slowly metabolized to
    formaldehyde and formic acid.
Health effects:
Acute: Headache, dizziness, dermatitis, conjunctivitis.
Chronic: Optic nerve damage and blindness.
Biological monitoring: Urine formic acid levels. Blood pH.
Treatment: Treatment of acidosis. Blindness is irreversible.
Measurement: Sampled on to a silica gel tube at an air-flow rate of
    50 ml min^{-1}, for subsequent analysis using gas chromatography. A
    general alcohol detector tube is available.
Control standards: HSE OES: 260 mg m^{-3} (200 p.p.m.) Sk.

Methyl bromide (bromomethane)

$$H-\overset{\overset{\displaystyle H}{|}}{\underset{\underset{\displaystyle H}{|}}{C}}-Br$$

Properties: Colourless, odourless gas. MW 92.95.
Uses: Fire extinguishers, refrigerant, insecticide and fumigant.
Metabolism: Rapidly absorbed by inhalation and toxic either directly or
    through metabolites such as bromide.
Health effects:
Acute: Late onset (several hours' delay) of acute respiratory tract irritation.
    Nausea, vomiting, headaches and convulsions may also occur.
Chronic: Recovery from the acute attack is usual but prolonged exposure

5 Chemical agents

5.3 Organic chemicals

---

or delayed treatment can cause peripheral neuropathy, tremor, renal failure and psychiatric disorders.

Biological monitoring: Non-specific.

Treatment: Non-specific. Treat convulsions and respiratory tract symptoms promptly.

Measurement: Sampled on to two large charcoal tubes in series at an air-flow rate of 1000 ml min^{-1}, for subsequent analysis using gas chromatography. Colorimetric detection tubes are also available.

Control standards: HSE OES: 20 mg m^{-3} (5 p.p.m.).

Methylene chloride (dichloromethane)

$$H-\overset{\displaystyle H}{\underset{\displaystyle Cl}{C}}-Cl$$

Properties: Non-flammable, colourless liquid. MW 84.93.

Uses: Paint and varnish remover, insecticide, fumigant, solvent, fire extinguisher.

Metabolism: Skin and lung absorption. Metabolized with production of carbon *monoxide.*

Health effects:

Acute: Skin and mucous membrane irritant. Can cause skin lesions. Acute intoxication with stupor, numbness and tingling of limbs.

Chronic: Dry, scaly dermatitis. Can precipitate cardiac insufficiency due to increase in carboxyhaemoglobin.

Biological monitoring: Carboxyhaemoglobin levels and methylene chloride concentrations in blood and exhaled air.

Treatment: Non-specific. Maintain adequate oxygenation.

Measurement: Sampled on to a charcoal tube at an air-flow rate of 1000 ml min^{-1}, for subsequent analysis using gas chromatography. Colorimetric detection tubes are also available.

Control standards: HSE MEL: 350 mg m^{-3} (100 p.p.m.).

Organophosphates

e.g.
$$(C_2H_5O)_2-\overset{\displaystyle }{\underset{\displaystyle S}{P}}-O-\!\!\left\langle\!\!\bigcirc\!\!\right\rangle\!\!-NO_2$$

A group of organics with phosphorus radicle, e.g. parathion.

Uses: Insecticide.

Metabolism: Inhibits cholinesterase activity leading to an accumulation of
endogenous acetylcholine in nerve tissue and effector organs.

Health effects:

Acute: Chest tightness, wheezing, blurred vision, increased salivation,
vomiting, diarrhoea, abdominal pain, bradycardia, frequency of urination,
fatigue. Death from asphyxia.

Chronic: Some organophosphates, including triorthocresylphosphate (TOCP)
can cause a delayed, sometimes irreversible, peripheral neuropathy.

Biological monitoring: Cholinesterase levels in red blood cells and serum.

*Treatment:*Atropine (i.v.), often in huge doses (2–4 mg m^{-1} at 5-min
intervals). Artificial respiration. Oximes (e.g. PAM) i.v.

Measurement:

Malathion or parathion: air is drawn through a glass fibre filter at a flow
rate of 1.5 l m^{-1}, for subsequent extraction using iso-octane for gas
chromatographic analysis.

Triorthocresyl phosphate: air is drawn through a cellular ester filter of pore
size 0.8 μm at a flow rate of 1.5 l m^{-1}, for subsequent extraction using
ether for gas chromatographic analysis.

Dichlorvos: air is drawn at a flow rate of 500–1000 ml m^{-1}, through a
tube containing XAD-2 resin, for subsequent desorption using toluene
for gas chromatographic analysis.

Control standards:

Dichlorvos: HSE OES: 1 mg m^{-3} (0.1 p.p.m.) Sk.

Malathion: HSE OES: 10 mg m^{-3} Sk.

Parathion: HSE OES: 0.1 mg m^{-3} Sk.

Tri-*o*-cresyl phosphate (tri-*o*-tolyl phosphate((TOCP): HSE OES: 0.1 mg m^{-3}.
MS 17.

Phenol

Properties: Colourless crystals. In solution, was used by Lister in his
historic carbolic sprays. MW 94.11.

Uses: Insecticides, disinfectants, pharmaceuticals, perfumes, explosives.

Metabolism: Readily absorbed by all routes. Oxidized to quinones and
excreted in urine, which darkens on standing due to formation of
homogentisic acid.

Health effects:
Acute: Powerful skin corrosive. Headache, dizziness, weakness, convulsions.
Chronic: Chronic dermatitis from low concentrations, severe scarring from phenol splashes. Renal failure. Weight loss, gastrointestinal disturbances.
Biological monitoring: Urinary phenolics.
Treatment: Liberal flushing with water of contaminated skin. Sedatives for convulsions. The use of polyethylene glycol eye irrigation has been advocated by some authorities.
Measurement: Air is drawn through a 0.1 N solution of sodium hydroxide in a bubbler at an air-flow rate of 1000 ml min^{-1}, an acidified solution is analysed using a gas chromatograph. Colorimetric detection tubes are also available.
Control standards: HSE OES: 19 mg m^{-3} (5 p.p.m.) Sk.

Styrene

Properties: Colourless liquid. MW 104.15.
Uses: Solvent for synthetic rubber, chemical intermediate, manufacture of polymerized synthetic.
Metabolism: Absorbed through lungs and skin. Rapidly metabolized to mandelic acid and, to a lesser extent, phenylglyoxylic acid and excreted in urine.
Health effects:
Acute: Acute mucous membrane irritation. Drowsiness, diminished cognitive and perceptual skills.
Chronic: Fissured dermatitis. Peripheral neuropathy (?). Long-term psychiatric sequelae (?).
Biological monitoring: Urinary mandelic acid and phenylglyoxylic acid concentrations.
Treatment: Non-specific.
Measurement: Sampled on to a charcoal tube at an air-flow rate of 200 ml min^{-1}, for subsequent analysis using gas chromatography.

5   Chemical agents

5.3   Organic chemicals

___

Control standards:
HSE MEL: 420 mg m^{-3} (100 p.p.m.).
MDHS 43 and 44.

Tetrachloroethane

$$Cl-\overset{\overset{H}{|}}{\underset{\underset{Cl}{|}}{C}}-\overset{\overset{H}{|}}{\underset{\underset{Cl}{|}}{C}}-Cl$$

Properties: Heavy non-flammable liquid. The most toxic of the chlorinated hydrocarbons. MW 167.85.

Uses: Solvent, artificial pearl production, dry-cleaning agent, chemical intermediate.

Metabolism: Rapid absorption from skin and lungs. Slowly metabolized and excreted in the urine. Main metabolite oxalic acid (?).

Health effects:

Acute: Gastrointestinal and upper respiratory tract irritation. CNS depression.

Chronic: Hepatic: hepatomegaly and hepatic failure. Neurological: polyneuropathy, particularly of the extremities. Haematological: mononucleosis. Renal: albuminuria. Dermatological: dry, scaly dermatitis.

Health surveillance and biological monitoring: Non-specific. Organ function tests may be valuable.

Treatment: Non-specific. Severe hepatic or neurological damage may be irreversible or even fatal.

Measurement: Sampled on to a charcoal tube at an air-flow rate of 200 ml min^{-1}, for subsequent analysis using gas chromatography.

Control standards: ACGIH TLV: 7 mg m^{-3} (1 p.p.m.) Sk.

Tetrachloroethylene (perchlorethylene)

$$\underset{Cl}{\overset{Cl}{>}}C=C\underset{Cl}{\overset{Cl}{<}}$$

Properties:

Non-flammable liquid with characteristic odour. MW 165.8.

Uses: Solvent widely used as dry-cleaning agent, fumigant.

Metabolism: Readily absorbed through the lungs and the skin.

5.3   Organic chemicals

Metabolized to trichloroacetic acid (TCA) and excreted in small amounts in the urine but mainly excreted unchanged in the breath.

Health effects:

Acute: Powerful narcotic. Can cause mucous membrane and skin irritation as well as liver damage.

Chronic: CNS depression and liver damage. Rodent carcinogen.

Biological monitoring: Blood TCA and breath analysis for original compound.

Treatment: Non-specific.

Measurement: Sampled on to a charcoal tube at an air flow rate of 1000 ml min^{-1}, for subsequent analysis using gas chromatography.

Control standards: HSE OES: 335 mg m^{-3} (50 p.p.m.). MDHS 28.

Toluene

Properties: Colourless liquid. MW 92.14.

Uses: Benzene manufacture, paint solvent, component of petrol.

Metabolism: Rapidly absorbed through lungs and skin and excreted as hippuric acid in the urine.

Health effects:

Acute: Narcotic. Conjunctival irritation and ulceration. Cardiac arrhythmias (has caused deaths in 'sniffers').

Chronic: Liver (?), kidney and bone marrow damage (compare benzene), but toluene exposure is rarely pure, exposure frequently including benzene ± xylene.

Biological monitoring: Urinary hippuric acid levels. A level of 5 g/l of urine correlated with an eight-hour time-weighted average of 200 p.p.m. of airborne toluene. Note: Hippuric acid is not a metabolite specific to toluene. It can be produced from dietary sources, such as food preserved with benzoic acid.

Treatment: Non-specific.

Measurement: Sampled on to a charcoal tube at an air-flow rate of 1000 ml min^{-1}, for subsequent analysis using gas chromatography. Colorimetric detection tubes are also available.

Control standards: HSE OES: 188 mg m^{-3} (50 p.p.m.) Sk. MDHS 40, 64 and 69.

5.3   Organic chemicals

1,1,1-Trichloroethane

$$H-\underset{\underset{H}{|}}{\overset{\overset{H}{|}}{C}}-\underset{\underset{Cl}{|}}{\overset{\overset{Cl}{|}}{C}}-Cl$$

Properties: Non-flammable liquid. MW 133.4.

Uses: Solvent, degreasing agent.

Metabolism: Readily absorbed through the lungs, and to some extent through the skin. Metabolized to chloroacetic acid and excreted in the urine.

Health effects:

Acute: Mucous membrane and skin irritant, narcotic, capable of sensitizing the myocardium to adrenaline.

Chronic: Dry, scaly dermatitis.

Biological monitoring: Urinary TCA estimations.

Treatment: Non-specific.

Measurement: Sampled on to a charcoal tube at an air-flow rate of 1000 ml min^{-1}, for subsequent analysis using gas chromatography.

Control standards: HSE MEL: 1900 mg m^{-3} (350 p.p.m.). MDHS 28.

Trichloroethylene

$$\underset{Cl}{\overset{H}{\diagdown}}C=C\underset{\diagdown Cl}{\overset{\diagup Cl}{}}$$

Properties: Non-inflammable liquid. MW 133.41.

Uses: Degreasing agent, anaesthetic gas.

Metabolism: Readily absorbed through the lungs and, to a lesser extent, the skin. Metabolized to chloral hydrate and then to trichloroacetic acid (TCA) or its glucuronide and excreted in the urine.

Health effects:

Acute: Powerful narcotic, action exacerbated by ethanol. Mild respiratory and skin irritant.

Chronic: Peripheral neuropathy has been reported. Addictive.

Biological monitoring: Urinary TCA estimations.

Treatment: Non-specific.

Measurement: Sampled on to a charcoal tube at an air-flow rate of 1000 ml min^{-1}, for subsequent analysis using gas chromatography.

5   Chemical agents

5.3   Organic chemicals

Control standards: HSE MEL: 535 mg m^{-3} (100 p.p.m.) Sk.
MDHS 28.

Trinitrotoluene

$$
\begin{array}{c}
CH_3 \\
NO_2 - \hspace{-0.5em} \bigotimes \hspace{-0.5em} - NO_2 \\
NO_2
\end{array}
$$

Properties: Colourless, explosive crystals. MW 227.14.
Uses: Explosives.
Metabolism: Absorbed through the skin and enhanced by sweating.
Health effects:
Acute: Cyanosis, mild anaemia. Irritant dermatitis and gastritis.
Chronic: Orange-stained skin. Toxic jaundice is rare, but when it occurs is
   frequently fatal. Aplastic anaemia.
Biological effect monitoring: Coproporphyrinuria is propotional to severity
   of poisoning.
Treatment: Thorough flushing of contaminated skin with water. The
   addition of potassium sulphate to the water is useful as contact with
   TNT produces a red colour. Otherwise, non-specific.
Measurement: Sampled on to a cellulose acetate filter of pore size 0.8 μm
   at an air-flow rate of around 2 l min^{-1}, for subsequent colorimetric
   analysis.
Control standards: HSE OES: 0.5 mg m^{-3}.

Vinyl chloride

$$
\begin{array}{ccc}
H & & H \\
 & C = C & \\
H & & Cl
\end{array}
$$

Properties: Flammable gas with pleasant odour. MW 62.5.
Uses: Polymerized to plastics, solvent in rubber manufacture. Previously
   used as aerosol propellant.
Metabolism: Rapidly absorbed by inhalation and partially excreted by the
   same route. Rapid clearance from blood through poorly understood
   metabolic pathways, but it is possible that some of the health effects

156

are related to the body reaction to vinyl chloride–protein complexes, which are considered 'foreign'.

Health effects:

Acute: narcotic.

Chronic: Fatigue, lassitude, abdominal pain. Raynaud's phenomenon, which can be severe. Acro-osteolysis leading to pseudo-clubbing and scleroderma-like changes. Angiosarcoma of the liver—rare but invariably fatal.

Health surveillance: Radiography of the hands. Liver function monitoring has been disappointing.

Treatment: Non-specific.

Measurement: Sampled on to two charcoal tubes in series at an air-flow rate of 50 ml min^{-1}, for subsequent analysis using gas chromatography. Colorimetric detector tubes are also available.

Control standards: HSE MEL: 3 p.p.m. averaged over one year. The annual maximum exposure limit is supplemented by an eight-hour TWA MEL of 7 p.p.m. for personal exposure with the proviso that the annual limit is not exceeded.

Xylene

ortho–   meta–   para–

Properties: Colourless liquid. MW 106.17.

Uses: Solvent. Chemical intermediate.

Metabolism: Rapidly absorbed from lungs, metabolized and excreted in urine as methyl hippuric acid.

Health effects:

Acute: Mucous membrane irritation. Narcotic.

Chronic: Aplastic anaemia has been postulated but may be due to benzene contamination (compare with toluene).

Biological monitoring: Urinary methyl hippuric acid.

Treatment: Non-specific.

Measurement: Sampled on to a charcoal tube at an air-flow rate of 1 l m^{-1}, for subsequent analysis using gas chromatography.

Control standards: HSE OES: 435 mg m^{-3} (100 p.p.m.) Sk.

5.4   Toxic gases

Although many organic compounds may be inhaled in vapour or gaseous form, the toxic gases *per se* are usually deemed to include compounds such as methane, sulphur dioxide and hydrocyanic acid.

In general, these gases may be classified as

- simple asphyxiants, e.g. nitrogen, carbon dioxide, methane
- chemical asphyxiants, e.g. carbon monoxide, hydrogen sulphide
- upper respiratory tract irritants, e.g. ammonia, sulphur dioxide
- lower respiratory tract irritants, e.g. oxides of nitrogen, phosgene.

Simple asphyxiants

These gases are only likely to be a danger when their concentration in the inhaled air is sufficient to cause a diminution in oxygen levels. Levels of oxygen below 14% lead to pulmonary hyperventilation and tissue anoxia.

Nitrogen

Nitrogen is the main constituent of air and is also present in high concentrations in some mines ('chokedamp'). Indeed the miners' canary and safety lamp were, in the main, introduced to detect such asphyxiating underground environments. In addition, nitrogen has industrial uses in ammonia production, as an inert atmosphere and as a freezing agent (boiling point $-195.8\,^{\circ}$C). In hyperbaric work such as diving, nitrogen becomes toxic, causing narcosis. It is normally detected chemically by eliminating other gases: what remains is assumed to be nitrogen.

Methane

Methane is the product of anaerobic decay of organic matter. Hence, it is found in sewers and wherever biodegradable organic matter is stored or dumped. It is also a natural constituent of fossil fuel reserves and is frequently found in coal mines and, occasionally, in other mines. Natural gas, used as a fuel in the UK, contains a large percentage of methane. It is explosive and lighter than air, and, therefore, in a concentrated form, will rise to make layers in unventilated ceilings and roofs which pose an explosive hazard. Explosive concentrations depend upon the percentage of oxygen present but, in fresh air, 5.2% is the lower explosive limit and 14% the upper explosive limit. Methods of detection started with the simple Davy lamp introduced in 1816, which, in modified form, is still used underground in coal mines today. The principle of detection involved a 'halo' of blue above the wick of a lowered flame, the shape of which indicates the concentration of methane. A double wire gauze prevents the

---

heat igniting methane outside the lamp. The latest instruments use solid-state sensors.

Carbon dioxide
Carbon dioxide occurs naturally as a product of combustion and of gradual oxidation and, hence, can occur wherever combustible or organic materials are to be found. Industrially, it is found as a by-product of brewing, coke ovens, blast furnaces and silage dumps. It has a wide use as an industrial gas, for example, in carbonization of drinks, brewing and refrigeration. It is heavier than air and in concentrated form, can produce 'pools' of inert atmosphere in low, unventilated places such as sumps and sewers. It occurs in mines in conjunction with nitrogen as a gas known as 'blackdamp' and in the aftermath of explosions as a gas known as 'afterdamp'.
Control standard: HSE OES: 9000 mg m^{-3} (5000 p.p.m.).

Carbon dioxide, unlike methane and nitrogen, is capable of stimulating the medullary respiratory centre to produce hyperpnoea. This begins to occur at a concentration of 3%, whilst, at 10% or more, loss of consciousness is rapid. It is also important to remember that, although the toxic effects of simple asphyxiants are readily reversed if removal from exposure and oxygenation is rapid, the 'weights' of the gases will determine the least hazardous approach by the rescue team to the stricken patient. Methane is lighter than air, carbon dioxide heavier, whilst nitrogen, comprising as it does 80% of normal air, is approximately the same density as air.

This characteristic of nitrogen, plus its inability to stimulate the respiratory centre, makes it a clinically inappropriate replacement for carbon dioxide as an inert gas for transportation of other products. Nevertheless, this is exactly what is happening in many industries today.

Chemical asphyxiants

Carbon monoxide (CO)
Occurrence: Produced by the incomplete combustion of carbonaceous compounds.
Properties: Colourless, odourless gas, burns with a blue flame. MW 28.0.
Uses: By-product of mining, smelting, petrochemical processes and many processes involving combustion.
Metabolism: High affinity of absorbed gas for haemoglobin, leading to

elevated carboxyhaemoglobin levels and diminished blood-oxygen carrying capacity. Excreted through the lung. Non-cumulative poison.

Health effects:

Acute: Insidious onset with giddiness, headache, chest tightness, nausea. Unconsciousness rapidly supervenes at concentrations in excess of 3500 p.p.m. No cyanosis (indeed, the patient (at post-mortem!) frequently has a deceptive healthy pink complexion due to carboxyhaemoglobin).

Chronic: Headache. Organic brain damage if asphyxiation was prolonged.

Biological monitoring: Carboxyhaemoglobin levels.

Treatment: Remove from exposure and give 95% O_2, 5% CO_2.

Measurement: Normally an immediate indication of concentration is required for safety reasons and, hence, it is measured by direct reading instruments, using a variety of principles. It can be sampled over a long period by slowly filling a container for subsequent analysis through a direct reading instrument. Colorimetric detector tubes are also available.

Control standards: HSE OES: TWA, 55 mg m^{-3} (50 p.p.m.). STEL, 330 mg m^{-3} (300 p.p.m.).

Legal requirements: Coal and Other Mines (Locomotives) Regulation, SI, 1956, No. 1771.

Hydrogen cyanide (HCN)

Occurrence: Gas emanates from contact of cyanide salts with acid.

Properties: Colourless gas with (apparently!) a bitter almonds smell.

Uses: By-product of metal manufacture, particularly the plating industry. Fumigant and steel hardener.

Metabolism: Inhibits action of cytochrome oxidase, thus disrupting oxygenation at tissue cell level.

Health effects:

Acute: Rapid onset of headache, hypopnoea, tachycardia, hypotension, convulsions and death. The rapidity of onset of symptoms necessitates treatment statim.

Chronic: None.

Biological monitoring: Blood cyanide concentrations.

Treatment: Remove contaminated clothing and wash skin. Administer amylnitrite inhalations and oxygen if the patient is breathing spontaneously. Dicobalt EDTA i.v. is advocated for the unconscious patient with a definite history of cyanide exposure. If in doubt, dispatch the patient immediately to hospital. Recent industrial experience suggests that the potentially toxic dicobalt EDTA need *not* be given immediately even in the unconscious patient unless vital signs are deteriorating.

Measurement: Sampled through a filter (to remove particulate cyanide interference) into a midget bubbler containing 0.1 mol l^{-1} potassium hydroxide at an air-flow rate of 2 l min^{-1}. The solution is analysed using a cyanide ion-selective electrode. Colorimetric detection tubes are also available.

Control standards: HSE MEL: STEL 10 mg m^{-3} (10 p.p.m.) Sk.

Hydrogen sulphide (H_2S)

Occurrence: Wherever sulphur and its compounds are worked.

Properties: Colourless gas with the smell of rotten eggs and a toxicity akin to hydrogen cyanide.

Uses: None of major importance.

Metabolism: Inhibits cytochrome oxidase (cf. HCN) and causes increase in sulphmethaemoglobin.

Health effects:

Acute: Lacrimation, photophobia and mucous membrane irritation in low concentrations. In high concentrations, paralysis of the respiratory centre can cause sudden unconsciousness.

Chronic: Keratitis. Skin vesicles. No cumulative effects.

Biological monitoring: None of great relevance.

Treatment: Removal from exposure. Administer oxygen and consider using amyl or sodium nitrite to convert haemoglobin to methaemoglobin, for combination with H_2S, thereby lowering effective H_2S concentrations.

Measurement: Sampled on to a molecular sieve tube via a desiccant tube of sodium sulphate at an air-flow rate of 150 ml min^{-1}, for subsequent analysis using gas chromatography. Colorimetric detection tubes are also available.

Control standards: HSE OES: 14 mg m^{-3} (10 p.p.m.).

Nickel carbonyl ($Ni(CO)_4$)

Occurrence: Generated during nickel refining (Mond process).

Properties: Colourless, odourless gas.

Uses: The unique properties of nickel carbonyl enable nickel to be extracted from the ore and subsequently released from the carbonyl gas in nearly 100% pure form.

Metabolism: Similar to carbon monoxide.

Health effects:

Acute: Headache, nausea, vomiting, unconsciousness. These symptoms

---

may subside and be followed up to 36 hours later with pulmonary
irritation and oedema.
Chronic: Nickel sensitivity. Chronic fibrotic lung disease (rare).
Biological monitoring: Carboxyhaemoglobin levels.
Treatment: Remove from exposure. Administer oxygen. Observe for at
   least 48 hours after initial exposure. Sodium diethyl-dithiocarbamate
   (Dithiocarb), orally or parenterally, depending on severity of symptoms.
Measurement: Sampled through an impinger containing a reagent at an air-
   flow rate of 2 l min^{-1}, for subsequent analysis using atomic adsorption
   spectrophotometry. Colorimetric detector tubes are also available.
Control standards: HSE OES: STEL 0.24 mg m^{-3} (0.1 p.p.m.).

Irritants

The irritant gases, as their name implies, are not respirable without
embarrassment. The somewhat artifical division into upper and lower
respiratory tract irritant is largely on the basis of solubility. Thus, the highly
soluble gases, such as ammonia, sulphur dioxide and chlorine, exert their
irritant effect on the upper respiratory tract which, unless the exposure is
prolonged and severe, saves the lungs. Conversely, gases of low solubility,
such as oxides of nitrogen and phosgene, have little effect on the upper
respiratory tract, their effect is delayed and the main brunt of the damage
is borne by the lungs.

Sulphur dioxide (SO_2)
Properties: Colourless gas with pungent odour and a density twice that
   of air.
Uses: Chemical and paper industries, bleaching, fumigation, refrigeration,
   preservative. A common by-product of smelting sulphide ores.
Metabolism: Products sulphurous acid on solution in water, leading to
   acidosis. Can be detected in concentrations as low as 3 p.p.m.
Health effects:
Acute: Acute mucous membrane irritant. The respiratory tract irritation is
   so severe that escape from the gas is imperative. Failure to escape
   leads to severe pulmonary oedema and death. Corneal ulceration and
   scarring.
Chronic: Diminution in olfactory and gustatory senses. Chronic bronchitis.
   Cataracts.
Bioligical monitoring: None of relevance.
Treatment: Remove from exposure, O_2, respiratory support.
Measurement: Sampled on to an impregnated cellulose filter, containing

potassium hydroxide, through a cellulose acetate pre-filter to collect particulate sulphates and sulphites, at an air-flow rate of 1.5 l min^{-1}. The impregnated filter is extracted with deionized water for subsequent anion exchange chromatography. Direct reading instruments and colorimetric detector tubes are also available and can also be sampled using a bubbler containing hydrogen peroxide for wet chemical analysis.
Control standards: HSE OES: 5 mg m^{-3} (2 p.p.m.).

Chlorine (Cl$_2$)
Properties: Greenish-yellow gas of pungent odour, over twice as heavy as air.
Uses: Chemical and pharmaceutical production, water disinfection, plastics manufacture.
Metabolism: Releases nascent oxygen from water and forms hydrochloric acid, resulting in severe protoplasmic damage.
Health effects:
Acute: Severe upper respiratory tract irritant leading to pulmonary oedema and death in those unable to escape its effects.
Chronic: Chronic bronchitis. Recovery from an acute exposure may be prolonged.
Biological monitoring: None of relevance.
Treatment: Remove from exposure. O$_2$ and respiratory support measures.
Measurement: Sampled by passing air through a fritted bubbler, containing 100 ml of dilute methyl orange at an air-flow rate of around 1.5 l min^{-1}, for subsequent colorimetric analysis. Colorimetric detector tubes are also available.
Control standards: HSE OES: 1.5 mg m^{-3} (0.5 p.p.m.).

Fluorine (F$_2$)
Properties: A greenish-yellow gas with a pungent odour. One of the most chemically active elements.
Uses: Fluorides are used as metal fluxes; uranium hexafluoride is used to separate isotopes of uranium. Glass etching, pottery, refrigeration (organic fluorides).
Metabolism: Converted to hydrofluoric acid in aqueous solutions.
Health effects:
Acute: Severe, penetrating, painful skin burns. Severe inhalational effects including laryngeal spasm, oedema and haemoptysis.
Chronic: Skin scarring. Pulmonary fibrosis. Fluorosis of the bones. Systemic effects of *hydrofluoric acid* exposure are related to the

disturbance of calcium and magnesium metabolism. Cardiac arrhythmias may follow lowered levels of these elements in the blood. Serum electrolyte estimation may indicate the need for calcium supplements.

Biological monitoring: Blood fluoride levels.

Treatment: Burns must be treated immediately with copious quantities of ice-cold water, followed by benzethonium chloride (for 2–4 hours) and then magnesium oxide/glycerine paste. Massaging calcium gluconate gel into the burn or infiltration of calcium gluconate solution at the site of the injury is considered by some to be the treatment of choice. Large burns require hospitalization and surgical débridement. Respiratory tract irritation requires intensive supportive measures.

Control standards: HSE OES: STEL 1.5 mg m^{-3} (1 p.p.m.).

Oxides of nitrogen (NO_x(N_2O, NO, NO_2))

Properties: NO_2 is the gas of greatest importance here and is reddish-brown with a pungent odour. Nitrous oxide (N_2O) is an anaesthetic gas.

Uses: Manufacture of nitric acid, explosives, jet fuel. It is generated during welding (some types), silo storage, blasting operations and diesel engine operation.

Health effects:

Acute: Insidious, due to slow progression of pulmonary irritation some 8–24 hours after exposure. Severe exposure can result in death from pulmonary oedema within 48 hours. NO_2 is the aetiological agent in silo-filler's disease.

Chronic: Brown discoloration of teeth. Transient patchy-lung opacities on chest radiography.

Biological monitoring: None of relevance.

Treatment: Non-specific. Observe all those exposed for at least 48 hours and admit to hospital anyone developing signs of respiratory irritation. Steroids and antibiotics may be necessary in the short term.

Measurement: Sampled on to impregnated molecular sieve tubes in tandem, at an air-flow rate of between 25 and 50 ml min^{-1}, for subsequent spectrophotometric analysis. Colorimetric detector tubes are available for nitrogen dioxide and nitrous fumes (NO + NO_2).

Control standards:

Nitric oxide (NO), HSE OES: 30 mg m^{-3} (25 p.p.m.).

Nitrogen dioxide (NO_2), HSE OES: 5 mg m^{-3} (3 p.p.m.).

Legal requirements: The Coal and Other Mines (Locomotives) Regulations, SI, 1956, 1771.

5 Chemical agents

5.4 Toxic gases

Phosgene (carbonyl chloride) $COCl_2$

Properties: Sweet smelling, highly toxic gas. MW 98.93.

Uses: Sources of chlorine, war gas. Evolution of phosgene is a hazard of burning chlorinated hydrocarbons, including many plastics.

Health effects:

Acute: Mild early symptoms followed by insidious onset of severe pulmonary oedema within succeeding 24–48 hours.

Chronic: No permanent lung damage in survivors.

Treatment: Hospital treatment of respiratory effects is obligatory.

Measurement: Air is drawn into a midget impinger containing nitrobenzylpyridine, at an air-flow rate of 1000 ml min^{-1}, for subsequent colorimetric analysis. Colorimetric detector tubes are available.

Control standards: HSE OES: 0.4 mg m^{-3} (0.1 p.p.m.).

Arsine (AsH_3), *phosphine* (PH_3) *and stibine* (SbH_3)

Arsenic, phosphorus and antimony are unique among the elements in producing hydride gases. Apart from the use of arsine in semi-conductor technology, all are of little or no commercial importance but are evolved when the elements are exposed to nascent hydrogen, as when metal dross is in contact with acidic water.

Arsine and stibine are both powerful haemolytic agents and can cause acute oliguric renal failure. Phosphine produces gastrointestinal and neurological symptoms. Long-term sequelae may be the result of the effects of the hydrides or the release of the elements themselves due to oxidation.

Measurement: Because of the acute nature of the toxic effects, emergency medication is normally required. Arsine is associated with a mild smell of garlic (odour threshold 0.5 p.p.m. or 10 times OEL) but stibine and phosphine odours cannot be described. Detection tubes are available for arsine and phosphine. Arsine can be collected on a charcoal tube at an air-flow rate of 200 ml min^{-1}, for subsequent analysis using atomic absorption spectrophotometry. Phosphine and stibine can be collected on an impregnated silica gel tube at an air-flow rate of 200 ml min^{-1}, for subsequent colorimetric analysis.

Control standards:

Arsine: HSE OES: 0.2 mg m^{-3} (0.05 p.p.m.).

Phosphine: HSE OES: STEL 0.4 mg m^{-3} (0.3 p.p.m.).

Stibine: HSE OES: as Sb, 0.5 mg m^{-3} (0.1 p.p.m.).

5   Chemical agents

5.4   Toxic gases

Legal requirements and HSE guidance:

Arsine (as a component of arsenic): The Factories (Notification of Diseases) Regulations, SI, 1966, No. 1400.

Arsine and phosphine (as compounds of arsenic and phosphorus): Factories Act, 1961, Section 82.

EH11—Arsine: Health and Safety Precautions.

EH12—Stibine: Health and Safety Precautions.

EH20—Phosphine: Health and Safety Precautions.

6 Physical agents

6.1   Introduction, 169

6.2   Particles and fibres, 169

6.3   Heat, 175

6.4   Noise, 188

6.5   Radiation, 198

6.6   Pressure, 202

6.7   Vibration, 203

167

6.1   Introduction
This chapter somewhat arbitrarily includes particles with the more 'traditional' physical agents such as heat, noise, radiation, pressure and vibration. The mechanical problems at work are highlighted in the section on 'repetitive strain disorders' (occupational overuse syndrome) in Section 8.6.

6.2   Particles and fibres

Coal dust (see also section on lung diseases, p. 91)
Occurrence: Mainly underground and world-wide. Formed due to the prehistoric accumulation of rotting vegetation.
Properties: Varies with the type and rank of coal.
Uses: Combustion, petrochemicals.
Metabolism: None.
Health effects:
Acute: None.
Chronic: Pulmonary fibrosis ranging from simple pneumoconiosis to progressive, massive fibrosis, which is a frequent precursor of death from respiratory failure.
Health surveillance: Lung function tests, especially spirometry. Serial chest radiography.
Treatment: Removal from exposure. Management of chronic respiratory disease.
Measurement: Sampled for airborne respirable fraction by drawing a known volume of air through a pre-weighted filter for reweighing and analysis. Respirable size selection for personal exposure is undertaken, by means of a cyclone separator at an air-flow rate of 2.0 l min^{-1} or, for static sampling, by means of a horizontal parallel plate elutriator, as in the MRE 113A sampler, at an air-flow rate of 2.5 l min^{-1}. As quartz dust is often found with coal, it may be necessary to determine respirable quartz content. Therefore, the following filters may be required:
for gravimetric analysis alone—glass fibre;
for X-ray diffraction analysis for silica—silver membrane;
for infra-red analysis for silica—PVC.
Control standards: Permitted levels of respirable dust in coal mines are laid down in the regulations given below. Exposure limits for workplaces other than coal mines: coal dust containing less than 5% quartz:

TWA 2 mg m^{-3} respirable dust; coal dust containing more than 5%
quartz: see silica.
Legal requirements:
Coal Mines (Respirable Dust) Regulations, SI, 1975.
Coal Mines (Respirable Dust Amendment) Regulations, SI, 1978.

Silica dust (see also section on lung disease, p. 91)
Occurrence: World-wide as earth's crust contains 28% silicon.
Properties: Silica (SiO$_2$) is a hard, rock-like compound capable of
    fragmentation into fine particles. It is a constituent of many ore-bearing
    rocks, coal seams, granites, china clay, sandstones and sand.
Uses: Abrasives, building materials, ceramics, foundry work, road stone.
Health effects: Inhalation into the lungs triggers a florid fibrotic reaction
    from the pulmonary tissues.
Acute: None.
Chronic: Severe nodular pulmonary fibrosis, mainly in the upper lung zones,
    with surface calcification of lymph nodes and a predisposition to
    tuberculosis. The possibility of a carcinogenic effect of silica on the
    lungs is discussed on p. 97.
Health surveillance: Pulmonary function tests, particularly spirometry. Serial
    chest radiography.
Treatment: Removal from exposure. Management of chronic pulmonary
    fibrosis.
Measurement: Airborne dusts that may contain silica require to be analysed
    for crystalline silica. It is usual to collect the respirable fraction by drawing
    a known volume of air through a pre-weighed filter for reweighing and
    analysis. Respirable size selection for personal exposure is undertaken by
    means of a cyclone separator at an air-flow rate of 2.0 l min^{-1} or, for
    static sampling, by means of a horizontal parallel plate elutriator, as in the
    MRE 113A sampler, at an air-flow rate of 2.5 l min^{-1}. The nature of the
    crystalline silica is determined by sampling on to a silver membrane filter
    and analysing, using X-ray diffraction analysis.
Control standards: See Table 6.1.
MDHS 51/2.

Asbestos (see also section on lung diseases, p. 91)
Occurrence: Naturally occurring fibrous silicates—either serpentine
    (chrysotile) or amphibole (crocidolite, amosite and anthophyllite).
Properties: Highly resistant to temperature, pressure and acids but these
    properties vary with the variety of asbestos. Serpentine varieties are

Table 6.1 Exposure limits for silica and other airborne dusts

	Respirable dust (mg m^{-3})	Total dust (respirable and non-respirable) (mg m^{-3})
Silica		
Crystalline quartz	0.1	0.3
Cristobalite	0.05	0.15
Tridymite	0.05	0.15
Tripoli	0.1	
Fused silica	0.1	
Mineral dusts		
Analyse for crystalline silica content and apply the above standards		
Amorphous silica	3.0	6.0
Talc	1.0	10.0
Non-siliceous dusts containing less than 1% quartz	5.0	10.0

also capable of being woven into cloth.

Uses: Many and varied, including asbestos cement, building and insulation materials, brake lining and fire-proofing devices.

Metabolism: Induces severe and possibly irreversible fibrosis in body tissues.

Health effects:

Acute: Nil of note.

Chronic: Chronic fibrotic lung disease—asbestosis. Pleural plaque formation and calcification. Carcinoma of the lung (synergistic effect with cigarette smoking). Malignant mesothelioma of pleura and peritoneum. Skin corns. Carcinoma of the larynx (?) and (possibly) ovary.

Health surveillance: Asbestos bodies in the sputum. Pulmonary function tests including spirometry and gas diffusion. Serial chest radiography.

Treatment: Removal from exposure. Management of chronic fibrotic lung disease and malignancies.

Measurement: It is necessary to determine the number of airborne respirable fibres by sampling on to a cellulose acetate filter for subsequent microscopic analysis and counting. A respirable fibre is defined as one that is greater than 5 μm in length and having a length:breadth ratio of at least 3:1 and a diameter less than 3 μm. Sampling should be done in accordance with HSE guidance note EH10 and MDHS 39/3.

6   Physical agents

6.2   Particles and fibres

Control standards:
HSE control limits:
(a) for asbestos consisting of or containing any crocidolite or amosite:
  (i) 0.2 fibres per millilitre of air averaged over any continuous period of
  4 hours;
  (ii) 0.6 fibres per millilitre of air averaged over any continuous period of
  10 minutes;
(b) for asbestos consisting of or containing other types of asbestos but not
crocidolite or amosite:
  (i) 0.5 fibres per millilitre of air averaged over any continuous period of
  4 hours;
  (ii) 1.5 fibres per millilitre of air averaged over any continuous period of
  10 minutes.
Legal requirements and HSE guidance:
The Control of Asbestos at Work Regulations, SI, 1987, No. 2115 as
  amended by SI, 1988, No. 712.
The Asbestos (Licensing) Regulations, SI, 1983, No. 1649.
The Asbestos (Prohibitions) Regulations, SI, 1985, No. 910.
Asbestos (Vol. 1) Final Report of the Advisory Committee (Simpson)
  HMSO, 1980.
Work with asbestos insulation and asbestos coating Approved Code of
  Practice and Guidance Note: revised edition, HSC, 1983.
A guide to the Asbestos (Licensing) Regulations HS(R) 19.
EH10 Asbestos—Control limits, measurement of airborne dust
  concentrations and the assessment of control measures (revised 1990).
EH35 Probable asbestos concentrations at construction processes 1989.
EH36 Work with asbestos cement 1989.
EH37 Work with asbestos insulation board 1989.
EH41 Respiratory protective equipment for use against asbestos 1985.
EH51 Enclosures provided for work with asbestos insulation and coating
  1989.
EH52 Removal techniques for asbestos insulation coatings and insulation
  1989.
MDHS 39/3 MS13.

Man-made mineral fibres (MMMF, synthetic mineral fibres)
Occurrence and uses: Manufactured by drawing, blowing, centrifuging and
  flame attenuation at very high temperature, using various raw materials
  and producing a range of fibres with differing diameters and properties.
Continuous filament: glass-drawn material, molten at 1000–1500°C,

3–15 μm diameter; product forms—yarn, roving, woven fabrics; uses—
    industrial textiles, glass-reinforced plastics.
Insulation wools: rockwool, slagwool, glass wool; basalt or dolerite rock,
    blast furnace slag, or glass-blown or centrifuged material at 1000–
    1500°C, 4–9 μm diameter; product forms—bulk fibre, blanket, slab,
    board, mattress; uses—acoustic and thermal insulation.
Refractory fibres: ceramic materials or alumina drawn or blown at
    1000–1500°C, 2–3 μm diameter; product forms—bulk fibre, needled
    blanket, board, paper, woven cloth, rope, specially moulded shapes;
    uses—high-temperature insulation for turbines, boilers, heat exchangers,
    fire protection, hot face linings in furnaces and kilns, seals, gaskets,
    expansion joints, high temperature gas and liquid filters. Have a low
    thermal mass, thus furnaces can be lighter and respond quicker.
Special purpose fibres: lime-free borosilicate glass flame-attenuated;
    0.1–3 μm diameter; product forms—bulk fibre, felt mat, blanket; uses—
    aero engine and rocket insulation, jet engine pipes and fuel systems.
Health effects: Fibres are not crystalline; they break transversely rather
    than longitudinally. Therefore, crushing and attrition does not make
    smaller fibres. The larger fibres cause skin irritation but there is no
    evidence of lung disease. Implantation and some inhalational studies on
    animals show that fibres below 0.25 μm diameter produce tumours.
    The production of fibres below 1 μm in diameter is a recent
    development and thus the effects on humans will not become apparent
    for many years. Recent international studies of MMMF workers have
    failed to resolve the question of whether MMM fibres are human
    carcinogens. At present, there remains a serious suspicion of
    carcinogenicity—particularly for the ceramic fibres and perhaps the
    rock/slagwool fibres. Glass fibres appear to be less implicated.
Measurements and standards: The HSE has published a gravimetric
    standard, that is, an MEL of 5 mg m^{-3} for fibres that are not respirable.
    Sampling is by drawing a known volume of air through a pre-weighed
    glass fibre filter mounted in a modified UKAEA sampling head; the
    weight gain in mg is divided by the volume of air sampled in m^3. For
    respirable fibres the method is similar to that of asbestos sampling and
    is explained in MDHS 59. A respirable fibre is defined as one that is
    greater than 5 μm in length and having a length:breadth ratio of at least
    3:1 and a diameter less than 3 μm.

Cotton dust (see also section on lung diseases, p. 91)
Occurrence: Cotton occurs on the plant as a 'boll' which is picked either by

hand or by machine but which usually contains some stem and leaves. The bolls pass through a 'ginning' process which separates the seeds and other materials from the raw cotton, which is then baled for transportation. The ginning process usually occurs close to the cotton fields. At the mills, the bales are opened and the cotton mechanically cleaned by blowing. It is then combed or 'carded' before being spun, dyed and woven. The dustiest areas are the blowing, blending and carding rooms, where the dust consists of 'fly', which is large fibres of cotton (up to 3 cm long), which are, however, too large to be inhaled, 'trash', which is a mixture of plant debris and soil, and fine, inhalable, cotton dust.

Health effects: The health hazard is from byssinosis, with its characteristic cough, chest tightness and difficulty in breathing, which is particularly prevalent on the first day back to work after a break, e.g. Monday mornings. Schilling has graded the symptoms of byssinosis into four grades from 0 to grade III.

Measurement and standards: Many studies have shown a direct relationship between total airborne dust measured in mg m^{-3} and prevalence of byssinosis. The standard is, therefore, gravimetric. The HSE OES for cotton dust is 0.5 mg m^{-3}, TWA of collected dust less fly. Sampling is, therefore, based on gravimetric techniques using a specially designed apparatus unique to the cotton industry. It draws air at about 10 l min^{-1}, through a glass fibre filter paper, which is screened by a gauze of 2 mm mesh, using 0.2 mm wire to eliminate the fly. During the sampling period, the screen has to be cleared of fly from time to time, to prevent it becoming a pre-filter reducing the collected fibre amounts.

Legal requirements and HSE guidance:

The Cotton Cloth Factories Regulations, S, R and O, 1929, No. 300. Cotton Cloth Factories Regulation Hygrometers Order, S, R and O, 1926, No. 1582.
EH25 Cotton dust sampling.
MS9 Byssinosis.

Dusts (other than those mentioned separately)
We are cautioned by the HSE that, although not all dusts have been assigned occupational exposure limits, the lack of such limits should not be taken to imply an absence of hazard, and that exposure should be controlled to the minimum that is reasonably practicable. However, where there is no indication of the need for a more stringent standard, personal

exposure should not exceed 10 mg m^{-3} total dust and
5 mg m^{-3} respirable (alveoli fraction) dust, except talc where the
recommended limit is: total, 10 mg m^{-3}, respirable, 1 mg m^{-3}. Any
airborne dust concentrations above these values should be regarded as
'substantial concentrations' for the purposes of Regulation 2 of COSHH
and, as such, are substances hazardous to health.
HSE guidance:
EH44 Dust in the workplace: general principles of protection.
MDHS 14.

Fume

This is regarded as solid particles generated by chemical reactions or
condensation from the gaseous state usually from the volatilization of
molten metal. Often the particles are in the region of 1 μm diameter unless
oxidization has taken place as with zinc fume, then diameters will be larger.

Control standards: HSE OES: welding fume, 5 mg m^{-3}.

6.3   Heat

Health effects

Body temperature is maintained within close limits by an efficient
homeostatic mechanism, though diurnal variation is observed over a range
of 0.5–1 °C. Physical exercise will increase body temperature in proportion
to oxygen consumption, the range being 0.5 °C, for moderate exercise, up
to 4 °C, for marathon running. In normal conditions, however, the body
temperature stays within the range 36–39 °C as a balance is struck
between the following:
• metabolic heat (M)
• evaporation (E)
• convection (C)
• conduction (K)
• radiation (R)
• storage (S).
Traditionally, this is expressed as

$$M = E \pm C \pm K \pm R \pm S$$

   Diving in a full suit or working in hot humid conditions can greatly alter
this homeostasis. For example, sweating ceases to be an effective means
of heat loss at ambient temperatures above 37 °C, at a relative humidity of

80%. Against this, is the fact that acclimatization to heat is possible over a period of 10 days and is facilitated by a greatly increased sweat rate. Furthermore, physical fitness improves an individual's ability to cope with the stresses of heat.

The severity of health effects from heat increases with the temperature, humidity and duration of exposure. In order of increasing seriousness, these effects are

- lassitude, irritability, discomfort
- lowered work performance and lack of concentration
- heat rashes
- heat cramps
- heat exhaustion
- heat stroke.

Any effect up to heat cramps is readily amenable to cooling and the administration of salt and water supplements. Heat exhaustion and heat stroke signify the onset of the failure of the thermoregulatory mechanism and this demands rapid and effective cooling, with fluid and electrolyte replacement by parenteral routes if necessary. Complete recovery of homeostasis may take a further week.

Environmental monitoring

The thermal environment around the body, which affects the rate of heat flow, is expressed by four parameters:

- the dry bulb temperature of the air
- the moisture content or water vapour pressure of the air
- the air velocity
- the radiant heat exchange between the skin and surrounding surfaces.

The relationship between the dry bulb temperature and the moisture content is shown in the psychrometric chart (Fig. 6.1). The two conditions which can be measured and plotted on this chart are the ventilated wet bulb and the dry bulb temperatures, as measured by the sling or aspirated psychrometer. Other factors, such as moisture content, percentage saturation (approximately the same as relative humidity), specific enthalpy and specific volume, can be read off on the appropriate scales from the point of intersection of the wet and dry bulb temperature, as shown in Fig. 6.2.

Air velocity is measured by an air-flow meter unless the value is low, in which case a kata thermometer is used. Air velocity is obtained from the cooling time of the kata thermometer, using the nomograms given in Figs. 6.3 and 6.4.

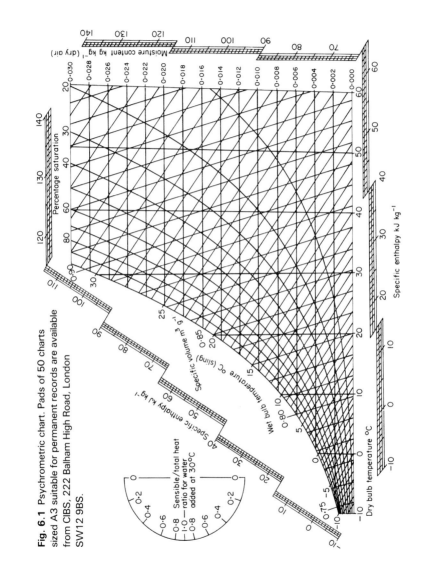

Fig. 6.1 Psychrometric chart. Pads of 50 charts sized A3 suitable for permanent records are available from CIBS, 222 Balham High Road, London SW12 9BS.

6.3   Heat

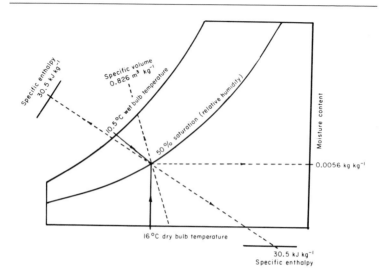

Fig. 6.2 Diagram to illustrate use of the psychrometric chart (from Ashton and Gill 1991). *Monitoring for Health Hazards at Work, 2nd edn,* Blackwell Scientific Publications, Oxford.

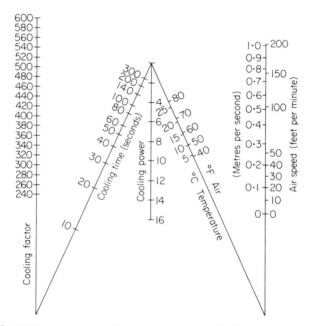

Fig. 6.3 Kata thermometer chart for temperature range 38–35°C (after a withdrawn British Standard, BS 3276, 1960).

6.3   Heat

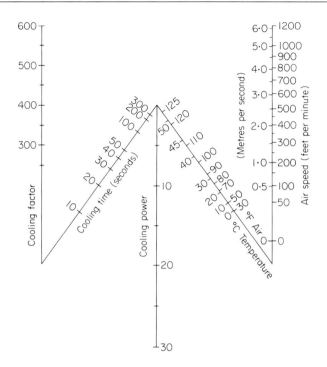

Fig. 6.4 Kata thermometer chart for temperature range 54.5–51.5°C (after a withdrawn British Standard, BS 3276, 1960).

Radiant heat exchange is obtained from the globe thermometer which integrates the radiant heat flux from all the surfaces which surround it. As the instrument is affected by air velocity, a correction is made using the nomograms given in Fig. 6.5a–d to provide the mean temperature of the surroundings (mean radiant temperature).

Other factors which affect body heat gain and losses are
• the metabolic rate of the subject due to degree of activity
• the type of clothing worn
• the duration of exposure to the heat or cold.

Typical *metabolic rates* for different activities are shown in Table 6.2 (p. 182).

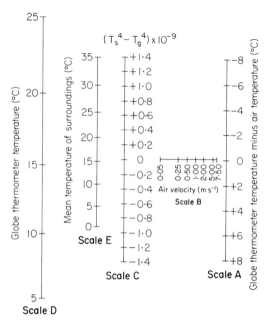

Fig. 6.5(a) Nomogram for the estimation of radiation from globe thermometer, range 5–25°C (after Waldron and Harrington (1980) *Occupational Hygiene*, Blackwell Scientific Publications, Oxford, p. 233).

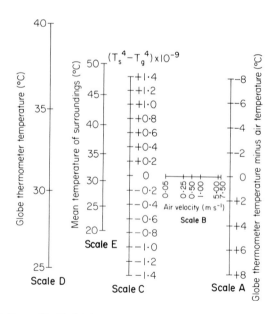

Fig. 6.5(b) Range 25–30°C (after Waldron and Harrington (1980), p. 234).

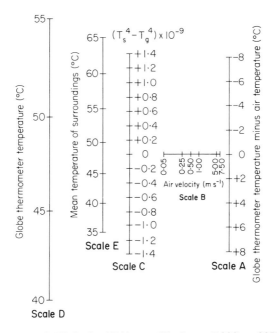

Fig. 6.5(c) Range 40–55 °C (after Waldron and Harrington (1980), p. 235).

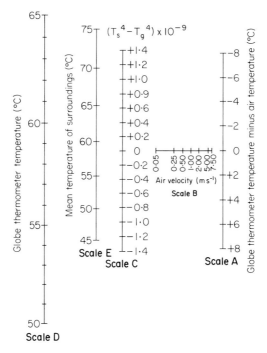

Fig. 6.5(d) Range 50–65 °C (after Waldron and Harrington (1980), p. 236).

6.3   Heat

Table 6.2   Typical metabolic rates

Acitivity	Metabolic rate (watts)
Sitting	95
Standing	115
Walking at 4 kr.i h^{-1}	260
Standing: light hand work	160–210
Standing: heavy hand work	210–260
Standing: light arm work	315
Standing: heavy arm work (e.g. sawing)	420–675
Work with whole body: light	315
Work with whole body: moderate	420
Work with whole body: heavy	560

Work rates tend to be self-regulating, in that a worker will voluntarily reduce his rate if he feels overheated except in fire-fighting and rescue work, where psychological pressures may overcome normal scruples.

Clothing assemblies have varying resistance to heat flow expressed by the unit 'Clo' (1 Clo $= 0.155°C$ m^{2}W^{-1}). Typical Clo values for various clothing assemblies are given in the following table.

Clothing assembly	Clo
Naked	0
Shorts	0.1
Light summer clothing	0.5
Typical indoor clothing	1.0
Heavy suit and underclothes	1.5
Polar clothing	3–4
Practical maximum	5

External factors such as moisture content and wind will influence the resistance of clothing to heat flow. Moist clothing will have a lower resitance. Higher air velocities tend to collapse clothing, reducing its thickness, and hence its resistance, whilst with open weave clothing wind can remove the inner layers of warm air. Except when used as a protection against chemicals or other hazards, personal insulation tends to be self-regulating, people adding or removing layers of clothing according to their feelings of comfort.

The *duration of exposure* can be varied by work/rest regimes, preferably with the rest period being taken in a less extreme environment. In certain circumstances, such as in hot mines and places of extreme climate, it may not be possible to remove the worker from the environment. This also occurs with rescue work, where the motive to continue the work at all costs is uppermost in the worker's mind.

Attempts have been made to bring together some of the factors mentioned above into a single index representing a thermal environment, from which the degree of hazard can be assessed. Some indices are given below.

Wet bulb globe temperature (WBGT)
For indoor use

WBGT $= 0.7\ t'_n + 0.3\ t_g$ °C

for outdoor use

WBGT $= 0.7\ t'_n + 0.2\ t_g + 0.1\ t$ °C

where t'_n = natural or unventilated wet bulb temperatures, t_g = globe temperature and t = dry bulb temperature.

Note that the ventilated wet bulb can be used instead of the natural wet bulb according to the following rules:
• if the relative humidity of the air is below 25% add 1°C to the ventilated wet bulb temperature
• if the relative humidity of the air is between 25 and 50%, add 0.5°C to the ventilated wet bulb temperature
• if the relative humidity of the air is above 50%, use the ventilated wet bulb temperature.

Recommended work/rest regimes for various WBGTs are given in Table 6.3.

Table 6.3 WBGT and recommended work/rest regimes

	Workload (total)		
	Light	Moderate	Heavy
Continuous	30	26.7	25.0
75% work, 25% rest each hour	30.6	28.0	25.9
50% work, 50% rest each hour	31.4	29.4	27.9
25% work, 75% rest each hour	32.2	31.1	30.0

Workload: Light 230 W; Moderate 230–400 W; Heavy 400–580 W.
For example: at a WBGT of 30°C a person could undertake continuous light work but if heavy work is involved s/he could only maintain it for 25% of the time in any hour.

6 Physical agents

6.3 Heat

Effective and corrected effective temperature (CET)
The three following charts in Fig. 6.6a–c give the basic normal and
adjusted scales of corrected effective temperature. The basic scale refers
to a worker stripped to the waist, the normal scale refers to a worker
lightly clothed and the adjusted scale takes into account the work rate.

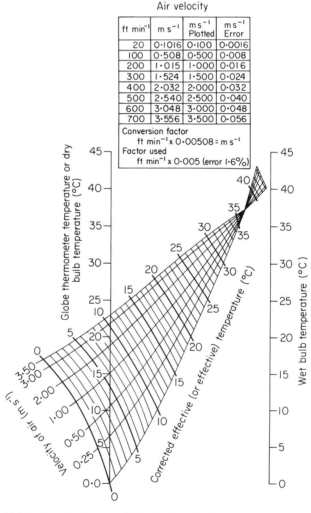

Fig. 6.6(a) Basic scale of corrected effective (or effective) temperature (stripped to
the waist) (after Waldron and Harrington (1980) *Occupational Hygiene*, Blackwell
Scientific Publications, Oxford, p. 241).

6.3   Heat

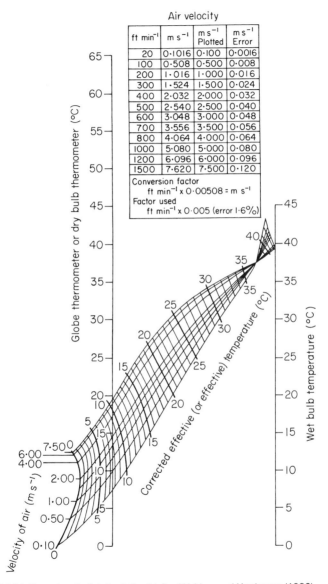

Fig. 6.6(b) Normal scale (lightly clothed) (after Waldron and Harrington (1980), p. 242).

6.3   Heat

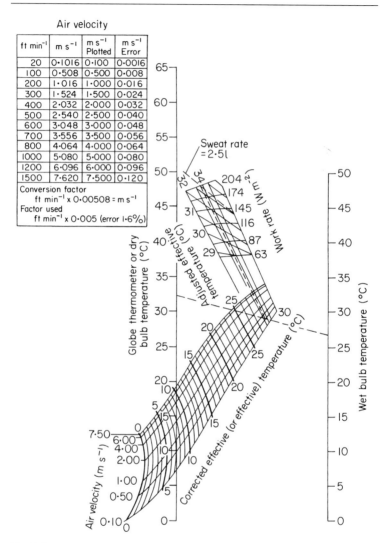

Fig. 6.6(c) Normal scale with additional nomogram including work rate (after Waldron and Harrington (1980), p. 243).

To use the charts, it is necessary to join the globe or dry bulb temperature reading to the wet bulb temperature with a straight line. The CET can be read from the nomogram, at the point of intersection of this line with the air velocity line.

6.3   Heat

Heat stress index (HSI)

This index is calculated as follows:

$$\text{HSI} = (E_{req}/E_{max}) \times 100\%$$

where $E_{req} = M + R + C$ watts.
For lightly clothed persons:

$E_{max} = 12.5\ v^{0.6}\ (56 - p_s)$ watts
M = metabolic rate of the worker in watts
$R = 7.93\ (t_r - 35)$ watts
$C = 8.1\ v^{0.6}\ (t - 35)$ watts

and for persons stripped to the waist:

$E_{max} = 21\ v^{0.6}\ (56 - p_s)$ watts
$R = 13.2\ (t_r - 35)$ watts
$C = 13.6\ v^{0.6}\ (t - 35)$ watts

where p_s = water vapour pressure in mbar, an approximate value of which can be found by reading the air moisture content in kg kg^{-1} from the psychrometric chart and multiplying that value by 1560; t_r = mean radiant temperature °C; t = dry bulb temperature °C; v = air velocity in m s^{-1}.

    A work/rest regime can be calculated from

$$\text{exposure time} = \frac{4400}{E_{req} - E_{max}}\ \text{min}$$

$$\text{rest time} = \frac{4400}{E_{max} - E_{req}}\ \text{min}$$

Note that, for the rest time, E_{max} and E_{req} refer to the thermal environment in the rest room, if one is used and normally the work rate will be less.

    The upper limit for safety is if the HSI reaches 100%. Any value above that will result in the deep body temperture rising, which, if allowed to continue for any length of time, may result in stress.

Wind chill index

In a cold environment the effect of wind is important. Figure 6.7 shows the equivalent still-air temperature of various wind velocities. The curves are labelled with a heat loss value in kcal h^{-1} m^{-2}. At a heat loss of 1750 kcal h^{-1} m^{-2} exposed flesh freezes in approximately 20 min but, at 2800 kcal h^{-1} m^{-2}, it freezes in 1 min.

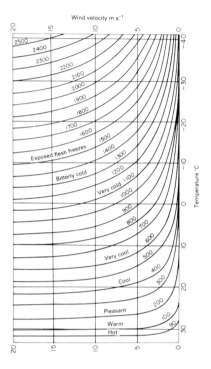

Fig. 6.7 Heat loss from body in kcal h^{-1} m^{-2}, for various air temperatures and wind velocities (after Waldron and Harrington (1980) *Occupational Hygiene*, Blackwell Scientific Publications, Oxford, p. 248).

6.4 Noise

Sound is pressure changes in the air which are picked up by the ear-drum and transmitted to the brain. Pressure is measured in pascals (Pa). The threshold of human hearing is at approximately 0.000 02 Pa but, at 25 m from a jet aircraft taking off, it is 10^7 times greater, at 200 Pa. Expression of such a wide range of sound is simplified with the decibel scale, which compares the actual sound with the reference value of 0.000 02 Pa, using a log scale (to base 10) as follows:

$$\text{decibel (dB)} = 20 \log_{10} \frac{p_a}{p_r}$$

where p_a is the pressure of the actual sound and p_r is the reference sound pressure at the threshold of hearing.

Typical sound intensities are given in Table 6.4.

6.4   Noise

Table 6.4 Typical noise levels

	Pressure (Pa)	Decibel (dB)
Threshold of hearing	0.000 02	0
Quiet office	0.002	40
Ringing alarm clock at 1 m	0.2	80
Ship's engine room	20	120
Turbo-jet engine at 25 m	200	140

Addition of sounds

If two sounds are being emitted at the same time, their total combined intensity is not the numerical sum of the decibel levels of each separate intensity. Because of the logarithmic nature of the decibel scale, they must be added according to the chart in Fig. 6.8.

Sound spectrum

The lowest frequency sound that can be detected by the human ear is at about 20 Hz and the highest, for a young person, is up to 18 kHz. With age, the ear becomes less sensitive to the higher frequencies. Doubling of the frequency raises the pitch of the note by one octave. The ear is most receptive to sounds between 500 Hz and 4 kHz, of which 500 Hz–2 kHz is the frequency range of speech. Unless a sound is a pure tone, which is unusual, most noises are made up of sounds of many frequencies and intensities and when assessing the intensity it may be necessary to discover what they are over the whole range of frequencies, that is, to

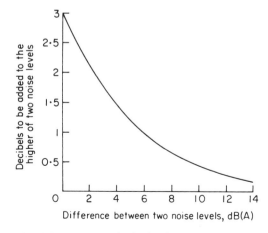

Fig. 6.8 Chart for adding two unequal noise levels.

measure the sound spectrum. For convenience, it is usual to divide the sounds into octave bands and to use a measuring instrument which assesses the intensities of all notes between the octaves and expresses it as a mid-octave intensity. The mid-octave frequencies chosen for this analysis are: 62.5 Hz, 125 Hz, 250 Hz, 500 Hz, 1 kHz, 2 kHz, 4 kHz, 8 kHz and, sometimes, 16 kHz. Thus a spectrum of noise will quote the intensities at each of these mid-octave band frequencies.

Noise rating (Fig 6.9)
Because of the sensitivity range of the ear, it can tolerate louder sounds at lower frequencies than at higher ones. A range of octave band curves

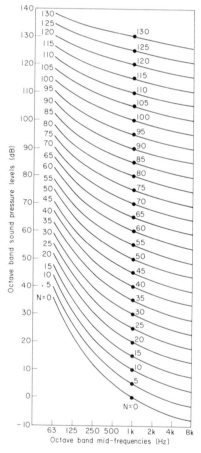

Fig. 6.9 Noise rating curves (after Waldron and Harrington (1980) *Occupational Hygiene*, Blackwell Scientific Publications, Oxford, p. 160).

known as 'noise rating (NR) curves' have been produced which indicate the recommended octave band analyses for various situations. The curve lying immediately above the measured octave band analysis of the noise in question represents the NR of that noise. A range of NR curves is given in Fig. 6.9 and the recommended rating for various situations in Table 6.5.

Table 6.5 Recommended noise ratings

Broadcasting or recording studio	NR15
Concert hall or a 500-seat theatre	NR20
Class room, music room, TV studio, large conference room, bedroom	NR25
Small conference room, library, church, cinema, courtroom	NR30
Private office	NR40
Restaurant	NR45
General office with typewriters	NR50
Workshop	NR65

Decibel weightings

As noise is a combination of sounds at various frequencies and intensities, the noise intensity can be either expressed as a spectrum, mentioned previously, or as a combination of all frequencies summed together in one value. As the human ear is more sensitive to certain frequencies than others, it is possible to make allowances for that in the electronic circuitry of a sound level meter. That is, certain frequencies are suppressed as others are boosted, in order to approximate to the response of the ear.

This technique is known as weighting, and there are A, B, C and D weightings available for various purposes. The one that is most usually quoted is the A weighting and instruments measuring sound intensity with that weighting give readings in dB(A). The weighting given to the mid-octave band frequencies for the dB(A) scale is given in Table 6.6.

Table 6.6 Mid-octave band frequency corrections for dB(A) weighting

Frequency (Hz)	62.5	125	250	500	1000	2000	4000	8000
Correction (dB)	-26	-16	-9	-3	0	$+1$	$+1$	-1

Noise dose and L_{eq}

Exposure to noise normally varies in intensity over a working period and, in order to estimate an equivalent noise level that would give the same total amount of sound energy as the fluctuating noise, the unit L_{eq} has been devised.

6.4   Noise

Table 6.7 Noise exposures equivalent to 90 dB(A) for 8 hours

Limiting dB(A) (L_{eq})	Maximum duration of exposure
90	8 h
93	4 h
96	2 h
99	1 h
102	30 min
105	15 min
108	7 min
111	4 min
114	2 min
117	1 min
120	30 s

A recommended maximum noise dose for the unprotected ear is given in Table 6.6. Each of the doses shown represents the same amount of sound energy and is regarded as 100% noise dose. Noise dosemeters sold on the British market are set to indicate a percentage dose based on this value. For example, if a dosemeter were placed on an unprotected worker for a period of time and the reading showed 150% then the recommended dose would have been exceeded but if it showed 30% then it would not.

This degree of energy does not fully protect some people from contracting noise-induced hearing loss if exposed day after day; a more suitable dose is 85 dB(A) for eight hours. A slightly modified version of these two noise doses is now written into the Noise at Work Regulations as the 'first action level' (85 dB(A) $L_{EP,d}$) and the 'second action level' (90 db(A) $L_{EP,d}$). The term $L_{EP,d}$ refers to a daily dose rather than one of eight hours. The Noise at Work Regulations are summarized in the next section. These regulations are published together with two HSE Noise Guides, and a further five Noise Guides are published separately. All are available from HMSO (see Chapter 13, Sources of Information).

Noise dose for workers exposed to reasonably steady sources of sound can be estimated from the chart in Fig. 6.10 taken from HSE Noise Guide 3. By drawing a line joining the duration of exposure on the right-hand side of the chart to the intensity of steady noise on the left-hand side, the $L_{EP,d}$ can be read from the centre line of the chart. Also shown on the centre line are numbers marked *f* which represents the fraction of 100% dose mentioned above. If a worker moves into other noisy areas, *f* numbers can be added from each to give a total fractional dose for the day.

6.4 Noise

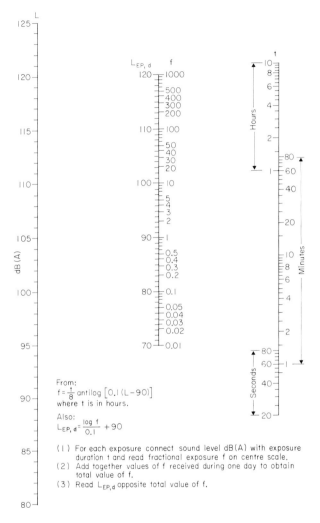

Fig. 6.10 Nomogram for the calculation of $L_{EP,d}$ (after Health and Safety Executive (1990) Noise Guide 3). Reproduced with permission of the Controller of Her Majesty's Stationery Office.

Example

An unprotected worker works in a steady 105 dB(A) for 10 minutes, 95 dB(A) for 4 hours and 88 dB(A) for 3 hours, What is his noise dose?

Answer:

105 dB(A) for 10 minutes	f = 0.7
95 dB(A) for 4 hours	f = 1.5
88 dB(A) for 3 hours	f = 0.25
Total	f = 2.45 or 245% dose

This level of noise is equivalent to a daily exposure of 94 dB(A) which is in excess of the 'second action level' as defined in the Noise at Work Regulations.

The Noise at Work Regulations 1989—SI No. 1790
The following is a summary of the regulations—the wording is that of the authors, not the HSC. They came into force on 1 January 1990.

Regulation 2—Interpretation
'First action level'—daily personal noise exposure of 85 dB(A).
'Second action level'—daily personal noise exposure of 90 dB(A).
'Peak action level'—sound pressure 200 Pa (140 dB).
Employer includes self-employed persons in respect of themselves.
Employee includes self-employed persons.

Regulation 3—Exceptions
On board ship under the direction of the master.
On board an aircraft or hovercraft under power.

Regulation 4—Assessment of exposure
Where any employees are above the first action level the employer shall have an assessment made by a competent person to
• identify which employees are so exposed
• obtain sufficient information to enable him to comply with Regulations 7, 8, 9 and 11.
   Review the assessment if
• it is no longer valid
• there is a significant change in the work.

Regulation 5—Assessment records
A record of that assessment or reviewed assessment shall be kept until a further noise assessment is made.

6 Physical agents

6.4 Noise

Regulation 6—Reduction of risk of hearing damage
Every employer shall reduce the risk of damage to the hearing of his/her employees from exposure to noise to the lowest level reasonably practicable.

Regulation 7—Reduction of noise exposure
When any employees are exposed to noise levels above the second action level or the peak action level the employer shall reduce the exposure so far as is reasonably practicable by means other than personal ear protectors.

Regulation 8—Ear protection
1 When employees are exposed to levels between the first and second action levels the employer shall provide the employee on request with suitable and efficient ear protectors.
2 When employees are exposed to levels above the second action level or the peak action level the employer shall take all reasonable steps to provide suitable ear protectors which, when properly worn, can reasonably be expected to keep the risks to hearing to below that level.

Regulation 9—Ear protection zones
Every employer shall ensure that
• ear protection zones are demarcated and identified by means of signs specificed in BS 5378 which show that it is an ear protection zone and that hearing defenders must be worn in that zone
• people who enter the zone must be wearing hearing defenders
• ear protection zone means wherever the employees are likely to be exposed to levels above the second action level.

Regulation 10—Maintenance and use of equipment
1 The employer must take reasonable steps to ensure that the equipment provided is (a) properly used and (b) maintained in an efficient state, in efficient working order and in good repair.
2 Every employee shall take all reasonable steps to make full and proper use of the equipment provided.

Regulation 11—Provision of information to employees
Every employer shall make adequate arrangements to provide each of his employees who is likely to be exposed to levels above the first action level or peak action levels adequate information, instruction and training on

---

* the risk of damage that such exposure may cause
* what steps the employee can take to minimize that risk
* how to obtain personal ear protectors
* the employee's obligations under these regulations.

Regulation 12—Modification of Section 6 of Health and Safety at Work Act regarding articles for use at work
Section 6 of HASAWA is modified to include a duty to ensure that articles used at work that are likely to result in an employee being exposed to the first action level or the peak action level are supplied with adequate information concerning the noise likely to be generated by them.

This regulation also applies to fairground equipment.

Regulation 13—Exemptions
Subject to a time limit and the power to revoke at any time the HSE may exempt any employer from

* Regulations 7 and 8(2), where the daily personal dose averaged over a week does not exceed 90 dB(A) and arrangements are made to ensure this
* Regulation 8(2) if compliance is likely to cause a risk to the health or safety of 8(2) and if it is not reasonably practicable to wear the ear defenders affording the highest degree of protection in the circumstances.

Such an exemption will not be granted if the HSE are of the opinion that the health and safety of persons are likely to be affected by it.

Regulation 14—Modifications for the Ministry of Defence
In the interests of national security the Secretary of State for Defence may exempt (subject to a time limit and powers to revoke)

* Her Majesty's Forces
* visiting forces or
* any member of a visiting force working in or attached to any headquarters or defence organization.

However, he must be satisfied that suitable arrangements have been made to assess and control the exposure to noise of these people.

Regulation 15—Revocation
Regulation 44 of the Woodworking Machines Regulations 1974 (a) is revoked.

Auditory health effects

The ear is not well equipped to protect itself from the deleterious effects of noise. Admittedly, a sudden loud sound is rapidly followed by a reflex contraction of muscles in the middle ear, which can limit the amount of sound energy transmitted to the inner ear. Nevertheless, in the occupational setting, such circumstances are relatively rare. Most workers exposed to noise suffer prolonged exposure, which may be intermittent or continuous. Such energy transmission, if sufficiently prolonged and intense, will damage the organ of Corti and eventually can lead to permanent deafness.

   Noise-induced hearing loss differs from presbycusis in being primarily centred on the ear's ability to hear sound at around 4 Hz—the upper level of speech appreciation. With time, this loss extends over the range 3–6 kHz and this has the effect of removing the sibilant consonants and, thereby, diminishing the hearer's appreciation of the spoken word. Unlike presbycusis, noise-induced hearing loss is not improved by the use of a hearing aid. The degree of hearing loss is related to the level of noise and the duration of exposure. Hence, the attempt to establish a maximum permissible exposure level for dB(A) L_{eq} with time, outlined above. Figure 6.11 shows progressive stages of noise-induced hearing loss.

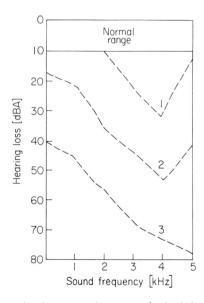

Fig. 6.11 Audiogram showing progressive stages of noise-induced hearing loss.

___

Hearing protection at work is clearly essential, either by lowering the ambient noise or by placing a barrier between the noise and the ears in the form of ear muffs, plugs or similar devices. Monitoring the effectiveness of this procedure usually means the use of audiometry.

Audiometry in industry serves several functions:
• the provision of a pre-placement baseline for the individual worker
• the provision of an opportunity for serial measures in an individual to follow his auditory status
• assistance in regulating the effectiveness of a hearing conservation programme for a group of workers
• demonstration to the worker of the effects of failure to wear hearing protection
• the identification of workers who are particularly susceptible to noise.

The procedure is simple to administer and painless to receive, though it must be done well. Certain specific questions are worth asking at each audiometric session. These relate to past auditory experience:
• ear pathology
• medical history
• occupational history
• military service
• hobbies.

It is, however, important to note that persons exposed to noise at work or at play are capable of exhibiting a temporary loss of hearing, which will recover when the noise ceases. This temporary threshold shift effect should preclude the administration of audiometry for 12 hours after cessation of exposure. Even so, there is some evidence that so-called permanent hearing loss is capable of minor reversal, though this is disputed.

Non-auditory health effects

High energy noise is also capable of producing visceral effects, such as heart rate alterations and changes in blood pressure and sweat rate. In addition, there are subtle psychosocial and psychomotor effects of trying to work in a noisy environment. It is, nevertheless, odd to visit a factory with moderate machinery noise levels and find that such environmental effects are upstaged by canned music blaring out at an even higher energy level!

6.5   Radiation

Radiation is energy which is transmitted, emitted or absorbed in the form of particles or waves. The effect of such radiation on living tissue is variable but the ability of this energy to ionize the target tissue distinguishes the two main sections of the electromagnetic spectrum, that is, ionizing radiation and non-ionizing radiation. The wavelength range of the whole spectrum is enormous: 10^{-4} to 10^{14}Å (10^{-12} cm to 1 km or more) (Fig. 6.12).

Ionizing radiation

The health effects of ionizing radiation can be divided into non-stochastic and stochastic ones, that is, respectively, effects for which there is a threshold (and thus a progression of severity of effect with dose) and no threshold. These are as follows:

Non-stochastic:

* acute radiation syndrome—gut, blood, CNS
* delayed—cataracts, dermatitis.

Stochastic:

* cancer, genetic damage.

Acute high-level exposure to, say, 20–50 Gy (2000–5000 rad) will result in death through cerebral oedema within 48 hours. Lower acute doses in the range of 1–20 Gy will also cause death through gastrointestinal or haematological damage, which may be delayed for up to one month.

Occupational exposures are at lower levels but are more prolonged. Thus, it is the neoplastic risk which is uppermost in most people's minds. The tumour site will depend on whether the source of radiation 'seeks' out a particular organ (e.g. ^{131}I and the thyroid, ^{90}Sr and bone) and on the site irradiated. Rapidly dividing cells are particularly radiosensitive, hence the bone marrow, gonads and gastrointestinal mucosa are at special risk. Additionally, care must be taken to protect target organs incapable or only slowly capable of repair (e.g. the ocular lens and nerve tissue).

Safe dose limits are recommended by the International Commission on Radiation Protection (ICRP) and have been adopted by the HSC as

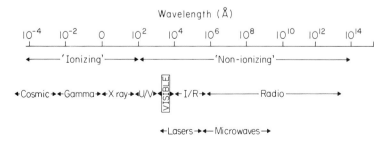

Fig. 6.12 Electromagnetic spectrum.

6.5   Radiation

published in the Ionizing Radiation Regulations 1985. The current dose
limits are given in Table 6.8.

To put this in perspective, consider the average person's dose in the
UK, (which is about 1m Sv). This is made up of the following radiation
components:

- cosmic 13%
- gamma 16% ⎤
- radon 33% ⎦ mainly building materials
- medical 20.7%
- fall-out 0.4%
- occupational 0.4%
- discharges 0.1%
- miscellaneous 0.4%

Occupational exposure is not, therefore, a major factor in 'normal'
radiation dosage but it is a potential danger for a large number of workers
(estimated to be in excess of 7 million workers in the USA).

Various instruments are available for radiological protection. They are
- ionization chambers
- Geiger–Müller tubes
- scintillation counters
- proportional counters
- film badges
- thermoluminescent dosemeters (TLD).

All the above are capable of being portable. Whereas the first four are
electronic, measure current radiation levels and are capable of giving

Table 6.8 Dose limits for ionizing radiation exposure which must not be exceeded in
a calender year (after Ashton and Gill (1991) *Monitoring Health Hazards at Work*,
Blackwell Scientific Publications, Oxford, p. 159).

	Employees aged 18 years or over	Trainees aged under 18 years	Any other person
Whole body	50 mSv (5.0 rem)	15 mSv (1.5 rem)	5 mSv (0.5 rem)
Individual organs and tissues	500 mSv (50 rem)	150 mSv (15 rem)	50 mSv (5 rem)
Lens of the eye	150 mSv (50 rem)	45 mSv (4.5 rem)	15 mSv (1.5 rem)

Women of reproductive capacity
Dose limit for the abdomen 13 mSv (1.3 rem) in any consecutive 3 month interval

Pregnant women
Dose limit during the declared term of pregnancy 10 mSv (1.0 rem)

audible and visible warning signals, the last two are measures of cumulative exposure and are the usual devices worn by radiation workers. No one instrument is capable of detecting or measuring all forms of radiation—α-, β-, γ- and X-rays. Most are usually designed to monitor the most penetrating radiations, such as γ- or X-rays.

Control can be summarized in three words:

• time
• distance
• shielding.

The first is obvious; the second is important because, for radiation, the inverse square law applies, that is, trebling the distance from the source reduces the radiation dose to one-tenth. Shielding will vary with the type of radiation source. Paper or water may be effective enough for α- and β-rays but thicker, denser, materials such as lead or concrete, may be required for γ-rays.

All that has gone before relates mainly to external sources of radiation. 'Internal' radiation must also be considered; thus it is important to make sure that the radiation source cannot penetrate the skin, be inhaled or be swallowed. Once inside, so to speak, removal of the source of radiation is a complicated and frequently incomplete procedure. Thus prevention of absorption is of paramount importance.

Legal requirements are listed below:

Radioactive Substances (Carriage by Road) (Great Britain) Regulations
    SI 1974/1735

Ionizing Radiations Regulations SI 1985/1333.

The protection of persons against ionizing radiations arising from any work
    activity—Approved Code of Practice.

In April 1991, the ICRP published revised recommendations for radiation protection. They call for a reduction in occupational exposure from 50 mSv in a year to an average of 20 mSv in a year over a 5 year period (100 mSv in 5 years), with a further provision that the dose limit should not exceed 50 mSv in any single year.

Non-ionizing radiation

For practical purposes, the two most important sources are lasers and microwaves. Both are capable of producing localized heating of tissues which may be intense and dangerous and both may be either continuous wave (CW) or pulsed.

The instrumentation required for measurement of *laser radiation* is complex due to the wide range of wavelengths and energies exhibited by commercial laser systems. Control measures largely revolve around instrument shielding and/or personal protective clothing, such as goggles. Measurement criteria for estimating maximum permissible exposure levels are given in the British Standards Institution Guide, BS 4803 (1983).

Microwave exposure limits vary in different countries. Table 6.9 provides a summary of those which have set standards. The devices used to monitor such radiations are usually small dipoles consisting of a diode or thermocouple device for converting microwave energy into electrical energy, which can then be measured by a voltmeter. Control measures to limit exposure consist of effective containment of the microwaves within the apparatus concerned. Mesh screens or even concrete may occasionally be used but personal protective clothing is rarely involved.

Table 6.9 Microwave exposure standards adopted by some countries (after Ashton and Gill (1991) *Monitoring Health Hazards at Work*, Blackwell Scientific Publications, Oxford, p. 188).

Country	Exposure ($W\,m^{-2}$)		Maximum
	Continuous	Intermittent	
UK	100	$10\ (0.1\ h^{-1})$	500
USA	100	$10\ (0.1\ h^{-1})$	250
Poland, Sweden Czechoslovakia and Canada	2	$Time = \dfrac{32}{(power\ surface\ density)^2}$ *	100
Russia	0.1	—	10

* Power surface density levels are set in units of watts per square metre.

6.6   Pressure

In normal circumstances humans are exposed to atmospheric pressure—variously defined in pounds per square inch, in millimetres of mercury, in SI units, newtons per square metre (pascals). Conveniently the pressure at sea level is 1 atmosphere, 1 bar, 14.2 pounds per square inch or 10^5 Pa ($N\,m^{-2}$). Rapid changes in pressure can be rapidly fatal but in occupational health the main problems concern

- deep sea divers
- caisson (tunnel) workers
- high-flying aviators.

The first two groups experience *increased* pressures, the last *decreased* pressures. Most of the ill effects of pressure in working environments result from the effects of decompression. These may be

- direct effects (barotrauma): ruptured tympanic membrane, ruptured

alveoli, sinus pain, dental cavity pain, etc.
• indirect effects due to nitrogen dissolved in the blood at *increased* pressure being released as bubbles on rapid decompression (remember Henry's Law?). This myriad of potential pathologies as nitrogen bubbles block blood vessels and damage tissues is called *decompression sickness*.

The symptoms of decompression sickness can be
• *acute*

Type I:	mild to severe limb pain
	skin mottling
Type II:	sensory or motor nerve effects to the limbs
	dizziness
	headache
	breathlessness
	chest pain
	convulsions
	coma

• *chronic*   permanent neurological or cerebral effects
                   aseptic bone necrosis.

The effects depend on the rate and degree of decompression and can range from minor discomfort to death. Treatment involves recompression in a special chamber and a measured decompression using the standard decompression tables (in the UK called the 'Blackpool tables', in the USA the 'State of Washington's tables').

The selection of workers for diving requires careful medical surveillance, which has to be carried out by approved medical practitioners. Particular attention is paid to dental and aural health, with the obese or those with previous cardiovascular or respiratory disease likely to be excluded from such work. Radiography of the chest and major joints are regularly undertaken. A previous history of Type II decompression sickness is also likely to preclude further hyperbaric work.

6.7   Vibration

Vibration is oscillatory movement about a point. In occupational health, workers may be exposed to two types:
• whole body
• hand–arm.

Vibration *magnitude* is expressed as root-mean-square acceleration in units of metres per second (m s^{-2} rms). Vibration *frequency* is expressed as cycles per second (Hz). Whole body vibration is usually in the range of 0.5–4.0 Hz and hand—arm 8–1000 Hz.

6.7   Vibration

Whole body vibration

This usually effects drivers of vehicles: tractors, coaches, helicopters—even ships. Specific organ effects depend on the natural resonance of that organ: thorax (3–6 Hz), head (20–30 Hz), jaw (100–150 Hz), and so on. In addition to the discomfort such organ 'shakes' can cause, longer-term effects leading to osteoarthritis of the spine have been described in some studies.

Hand–arm vibration

This is an occupational hazard of

• chain saw operators
• fettlers
• power grinders
• hammer drillers

The effects are easier to describe than the pathophysiology is to explain. These effects are called the hand–arm vibration syndrome (HAVS) and constitute)

• vascular effects—episodic blanching of terminal digits exacerbated by cold temperatures (Raynaud's phenomenon)
• neurological effects—numbness and tingling of terminal digits.

The effects are progressive with continued exposure to vibrating tools and can lead, in the worse cases, to gangrene.

Various classifications and staging of HAVS have been formulated. The latest is the Stockholm modification (1987) of the Taylor and Pelmear Scale (1974) (Table 6.10). In addition it is useful to score the extent of phalangeal involvement using a scale proposed by Griffin (1982). Blanching, numbness, tingling and colour changes may be scored separately.

Unfortunately, there is no single definitive test for HAVS. The following are some of those in vogue:

• cold provocation
• photoplethysmography
• digital blood pressure (Doppler)
• Dermal sensitivity (aesthesiometry).

Limitation of exposure to vibration is not easy, given that many of the tasks require the operators to be able to 'feel' what they are doing. The measures include

• control vibration at source—accelerometers, tachometers
• isolation—springs, mountings, dampers.

6.7   Vibration

Table 6.10 Two-tier 'Stockholm classification' for hand–arm vibration syndrome
(a) Cold-induced Raynaud's phenomenon staging

Stage*	Grade	Description
0		No attacks
1	Mild	Occasional attacks affecting only the tips of one or more fingers
2	Moderate	Occasional attacks affect distal and middle (rarely also proximal) phalanges of one or more fingers
3	Severe	Frequent attacks affecting all phalanges of most fingers
4	Very severe	As in stage 3, with trophic skin changes of most fingers

* The staging is made separately for each hand. In the evaluation of the subject, the grade of the disorder is indicated by the stages of *both hands* and the number of affected fingers on each hand; example: '2L(2)/1R(1)', '−/3R(4)', etc.

(b) Sensorineural staging

Stage*	Symptoms
OSN	Exposed to vibration but no symptoms
1SN	Intermittent numbness, with or without tingling
2SN	Intermittent or persistent numbness, reduced sensory perception
3SN	Intermittent or persistent numbness, reduced tactile discrimination and/or manipulative dexterity

* The sensorineural stage is to be established for each hand.

British (BSI) and international (ISO) standards exist to provide guidance on 'acceptable' vibration exposure. For many purposes, the 8-hour equivalent magnitude of 2.8 m s^{-2} rms can be considered an appropriate action level above which preventive measures are necessary.

7 Biological agents

7.1 Ankylostomiasis, 209

7.2 Anthrax, 209

7.3 Brucellosis, 210

7.4 Glanders, 210

7.5 Serum hepatitis (hepatitis B, HBV), 211

7.6 Leptospirosis, 211

7.7 Malaria, 212

7.8 Q fever, 212

7.9 Psittacosis, 213

7.10 Tuberculosis, 213

7.11 Legionnaire's disease, 213

7.12 Expatriates' health, 214

7   Biological agents

Biological hazards at work include
- acute and chronic infections
- parasites
- toxic products
- allergic reactions to plant and animal agents
- irritants.

Most manual workers are prone to cuts and abrasions, which may be infected. In this section, specific reference will be made to the more notable infectious agents which can cause occupational diseases. In the main, these are the zoonoses, and contact with animals or animal products is, thus, a common thread connecting many of the diseases cited below. Biologically active dusts which cause extrinsic allergic alveolitis are described on p. 99.

It is important to note that travel forms an important part of the occupational activities of many people these days, be they engineers, executives or manual workers. It is, therefore, possible to subsume almost any tropical infectious disease under the heading of occupationally related disorders!

7.1   Ankylostomiasis (hookworm)

Occurrence: Tropical countries, particularly with warm, moist, sandy soil. Can be found in mines in temperate zones.

Aetiological agent: Helminths—*Necator americanus* and *Ankylostoma duodenale.*

Exposures: Mining, agriculture, construction sites in tropical countries.

Mode of transmission: Eggs in faeces are deposited on ground and larvae can penetrate intact human skin. These pass via lymphatics and blood, to the lungs, trachea, oesophagus and gastrointestinal tract, where they attach, mature and shed further eggs.

Health effects:

Acute: Severe pruritis at site of larval entry.

Chronic: Non-specific gastrointestinal symptoms and microscopic bleeding. Iron-deficiency anaemia.

Diagnosis: Eggs in faeces. Microscopy of adult worms for species recognition.

Treatment: Hexylresorcinol or bephenium hydroxynaphthoate.

7.2   Anthrax

Occurrence: Infrequent and sporadic in Western countries. Endemic in agricultural regions of underdeveloped areas of the world.

7.3   Brucellosis

Aetiological agent: Extremely resistant, Gram-positive, spore-forming
   bacillus.

Exposures: Workers handling animals and animal products, especially
   unsterilized hair, bone and hide.

Mode of transmission: Skin infection by contact. Pulmonary infection by
   inhalation of spores. Gastrointestinal infections by ingestion (rare).
   Incubation period, 2–3 days.

Health effects: Cutaneous eschar: pustule with ring of blisters and marked
   oedema. Pulmonary: non-specific symptoms for a few days, followed
   by fever, respiratory distress and death within 24 hours.

Diagnosis: Culture of pustule.

Treatment: Penicillin (in large doses). Immunization is now available for
   workers at risk.

7.3   Brucellosis

Occurence: World-wide.

Aetiological agent: Gram-negative bacteria—*Brucella abortus* (cattle),
   B. melitensis (sheep and goats), *B. suis* (pigs).

Exposures: Slaughter houses, veterinary surgeons, farmers, laboratory
   workers.

Mode of transmission: Direct contact with abraded, broken skin and
   conjunctiva. Incubation period, 5–21 days.

Health effects:

Acute: Fever, headaches, arthralgia, low back pain. Lymphadenopathy and
   (possibly) splenomegaly. Recovery normally within two weeks, unless
   complications occur, such as endocarditis, spondylitis, liver and splenic
   disorders.

Chronic: A disputed entity. Vague, non-specific disorders including
   lassitude, debility, low-grade fever, neurasthenia.

Diagnosis: Blood culture.

Treatment: Tetracyclines.

7.4   Glanders

Occurrence: Largely eliminated throughout the world. Enzootic foci in
   Mexico and Mongolia.

Aetiological agent: Gram-negative organism (*Actinobacillus mallei*).

Exposures: Working with horses, mules and donkeys.

Mode of transmission: Contact.

Health effects:

Acute: Acute, severe lymphangitis with suppuration and fever.

Chronic: Painful ulcers at site of original infection—usually the skin, rarely the lungs.

Diagnosis: Culture.

Treatment: Sulphonamides, streptomycin, tetracycline or chloramphenicol. Untreated, the fatality rate exceeds 90%.

7.5   Serum hepatitis (hepatitis B, HBV)

Occurrence: World-wide, endemic with seasonal variation.

Aetiological agent: A DNA virus.

Exposures: Medical and paramedical staff, especially in renal dialysis units, blood transfusion centres, venereal disease clinics and laboratories.

Mode of transmission: Parenteral, by inoculation usually. Oral and venereal spread have been postulated. Incubation period, 80–100 days.

Health effects:

Acute: malaise, myalgia, headache, fever ± jaundice. Can be fatal and it is usually more severe than infective hepatitis (hepatitis A, HAV).

Chronic: Chronic active hepatitis, increased risk of hepatocellular carcinoma, and/or superinfection with delta agent hepatitis (HDV).

Diagnosis: Liver function tests. The presence of surface hepatitis B antigen in the blood.

Treatment: No specific treatment. There is now an effective hepatitis B vaccine.

Note: Hepatitis C (HCV) is also spread parenterally and is therefore a potential occupationally acquired infection.

7.6   Leptospirosis

Occurrence: World-wide, associated with water.

Aetiological agent: Motile spirochaete-like organisms including *Leptospira icterohaemorrhagica, L. canicola* and *L. hebdomadis.*

Exposures: Sewer and canal workers, swimmers, farmers, paddy field workers, slaughterhousemen, veterinarians.

Mode of transmission: Penetration of abraded skin or intact conjuctiva. Incubation period, 4–19 days.

Health effects:

Acute: fever, headache, malaise, vomiting, myalgia, meningism (occasionally); hepatic or renal complications (rarely).

Diagnosis: Leptospiraemia (1st week). Albuminuria (2nd week). Leptospiruria (3rd week).

Treatment: Penicillin.

7.7   Malaria

Occurrence: Mainly tropical and subtropical regions.

Aetiological agent: Various plasmodia— parasitic organisms with asexual phase in erythrocyte and sexual phase in anopheline mosquito.

Exposures: Living in tropics, agricultural advisers, temperate zone expatriates and travellers.

Mode of transmission: Bite from infected anopheline mosquito.

Health effects:

Acute: (10 + days after bite depending on species), fever with daily variation, rigors, delirium, haemolytic anaemia. N.B. *Plasmodium falciparum* malaria is very dangerous, due to cerebral damage leading to coma and death.

Chronic: Recurrent bouts of fever for many years if inadequate treatment allows persistence of intrahepatic phase of life cycle. Splenomegaly and hepatomegaly.

Diagnosis: Parasitaemia.

Prophylaxis: Several choices—need to be started well before exposure and to be continued for some time after leaving malarious area.

proguanil—daily

pyrimethamine—weekly

chloroquine—weekly

pyrimethamine/sulphadoxin—weekly.

A vaccine is undergoing human trials.

Treatment: Always check local resistance, especially in South East Asia and East Africa. 4-Aminoquinoline, chloroquine, amodiaquine.

7.8   Q fever

Occurrence: Endemic in Australia (Queensland, hence the 'Q'), USA and elsewhere.

Aetiological agent: Rickettsia—*Coxiella burneti.*

Exposures: Farmers, veterinarians, laboratory workers, wool processors. Incubation period 2—3 weeks.

Mode of transmission: Tick-borne or airborne dissemination of rickettsia in dust or birth fluids. Direct contact with infected wool, straw, milk.

Health effects:

Acute: influenza-like illness. Patchy consolidation of the lungs. Meningism.

Chronic: Endocarditis has been reported. Unapparent infections are possible.

Diagnosis: Antibody titres in serum.

Treatment: Tetracycline or co-trimoxazole.

7.9 Psittacosis
Occurrence: World-wide.
Aetiological agent: Chlamydia organism.
Exposures: Mainly confined to occupations associated with psittacine
  birds, which may themselves be healthy carriers.
Mode of transmission: Inhalation of the agent in the dust from desiccated
  bird droppings. Incubation period 4–15 days.
Health effects:
Acute: Acute generalized infectious disease, with fever, headache and
  pneumonitis. Can be severe, but is rarely fatal.
Chronic: Relapses are common.
Diagnosis: Antibody titres in serum.
Treatment: Tetracycline.

7.10 Tuberculosis
Occurrence: World-wide.
Aetiological agent: Mycobacterium, usually *M. tuberculosis.* Gram-
  negative, acid-fast bacillus.
Exposures: Medical and paramedical staff. Agricultural workers and
  veterinarians occasionally acquire *M. bovis.*
Mode of transmission: Inhalation of airborne organisms. Ingestion (rarely).
  Incubation period, 4–6 weeks.
Health effects:
Acute: Primary infection usually unapparent.
Chronic: Reactivation of latent focus with chronic, variable course of
  pulmonary infiltration, cavitation and fibrosis. Cough, night sweats,
  weight loss, chest pain and haemoptysis frequently occur.
Diagnosis: Sputum culture.
Treatment: Various effective therapies of single or combination drugs,
  including streptomycin, isoniazide, *para*-aminosalicylic acid and
  rifampicin.

7.11 Legionnaire's disease
Although not strictly an 'occupational disease', this infection does have
occupational connotations. Legionnaire's disease is a form of pneumonia
due to *Legionella pneumophila.* The name originates from an outbreak
described in a group of participants attending an American Legion
convention at a hotel in Philadelphia in 1976.

7.12 Expatriates' health

Causative agent
Ubiquitous aerobic Gram-negative flagellate bacillus. It is found in nature in soil, pools of water, and in water samples from cooling towers. It grows optimally at 35°C (range: 20–45°C). The organism is transmitted by airborne particulates rather than by person-to-person spread.

Clinical features of disease
Legionnaire's disease tends to affect susceptible individuals such as immunocompromised patients in hospitals, or the aged, especially those with chronic pulmonary disease. Affected patients present with fever, non-productive cough, headache and malaise. Other features can include respiratory, gastrointestinal, cardiovascular and central nervous system symptoms. Chest X-rays show patchy consolidation. Diagnosis is on the basis of a fourfold or greater increase in antibody titre. Erythromycin is the treatment of choice.

Pontiac fever is a limited non-pneumonic form of infection by *Legionella*, with a shorter incubation period.

Occupational health perspective
Staff exposed to the organism may develop antibodies without clinical disease. Workplace exposures include buildings where the bacilli proliferate in the water and spread through the ventilation or water supply system. Outbreaks have been reported from hospital cooling towers and in hotels, office blocks and large commercial buildings. Hospital inpatients and hospital visitors and outpatients are amongst the groups affected.

Control measures
These include cleaning up ventilation, water supply and water humidification systems, keeping the water supply below 20°C or above 45°C, and hyperchlorination of water.

Biocides can be used, though with care as they can affect some exposed individuals.

Looking for the organism in water supply or ventilation systems in the absence of an outbreak is not recommended, as the bacillus is widespread in nature.

7.12 Expatriates' health
Many occupational health practitioners are required to provide advice to employees travelling abroad on company business. Many large organizations supply their own travel kits (containing items like analgesics,

anti-malarials, anti-diarrhoeals, etc., as well as advice on what to eat and drink). Some even have their own vaccination/immunization service.

The main hazards facing the traveller are endemic and epidemic infectious diseases to which s/he has little or no immunity. The most important infections are
• blood-borne viruses—HBV, AIDS
• sexually transmitted disease—including HBV and AIDS
• brucellosis
• diarrhoeal disease: dysentery (amoebic and bacillary); typhoid fever; giardiasis; hepatitis A
• malaria
• rabies
• schistosomiasis
• arboviruses, e.g. Japanese B encephalitis, dengue, haemorrhagic fever and yellow fever (West and Central Africa, Central and South America).

In addition all travellers to countries other than Europe, North American and Australasia should ensure that their polio and tetanus protection is up to date.

Travellers who return home seeking medical advice usually have one or more of the following symptoms:
• fever (real or imagined)
• diarrhoea
• jaundice
• anxiety about the possible acquisition of an exotic disease.

8 Special issues in occupational health

8.1 Introduction, 219

8.2 Accidents, 219

8.3 The health of health-care workers, 219

8.4 Medical audit, 223

8.5 Mental health at work, 224

8.6 Occupational cancer, 226

8.7 Repetitive strain injury (RSI), 232

8.8 Rehabilitation and resettlement, 234

8.1   Introduction
In addition to the occupational diseases presented in the previous
chapters, the occupational health practitioner is faced with other problems
which cannot always be neatly classified in section headings. What follows
is a series of short accounts on aspects of occupational health of current
interest or concern.

8.2   Accidents
Accidents at work are an important cause of personal suffering and
lowered productivity. In the United Kingdom, the latest figures available are
for 1988/89 when major injuries (3 days +) totalled 196 499 with 730
deaths. The deaths include 167 from the Piper Alpha disaster. Excluding
Piper Alpha, the fatality rate is relatively stable at 1.7 per 100 000
employees with the highest rate (9.9) for the construction industry. The
injury rate averages 0.8 per 100 000 employees but is much higher in
construction, manufacturing and extractive industries, with the highest rate
(6.2) for coal extraction.

   The causes of accidents are difficult to summarize. There have been
vogues for suggesting that it is all due to 'accident proneness' or at the
other extreme it is all due to 'chance'. Many accidents are clearly
multifactorial. Some of the more important factors are

- age
- experience
- time of day
- work rate
- type of work
- health of the worker
- industrial relations

   Accident prevention must therefore be based on a knowledge of these
factors with good planning for safety, a balanced package of control
measures and the availability of trained personnel to design and supervise
such measures.

8.3   The health of health-care workers
Health-care workers comprise a large workforce, and in the UK form the
largest group of employees in any industry. Workers in the health-care
industry extend beyond the different categories of doctors and nurses
representing various specialities from hospital-based surgical and medical
staff to community nurses and family physicians. Para-medical staff such
as ambulance crew, physiotherapists and laboratory technicians, and

administrative and support staff such as hospital engineers and kitchen and catering personnel are all employees within the health-care industry.

Renewed interest in the occupational health status of health-care workers has come with the removal of Crown immunity and with the realization that the abundance of doctors and nurses in the health service does not necessarily protect against occupational hazards in that environment. Occupational health departments in the NHS (National Health Service) are staffed mainly by occupational health nurses although the number of consultant occupational physicians in the NHS has shown a steady increase from a handful five years age to more than 50 in 1991. However, there are still only three full-time occupational hygienists employed in occupational health departments in the NHS in the UK. With the current organizational and administrative changes in the health service, it is difficult to predict the trend and emphasis in occupational health staff numbers and functions in the NHS.

Some examples of specific occupational hazards

Physical
• Heat—as in underground tunnels for hospital pipes and cables. Laundry facilities, boiler rooms and kitchens are other areas where high workplace temperatures may be encountered.
• Noise—can be high in hospital engineering workshops. Concerns have also been raised about tinnitus and auditory effects from occupational exposure to ultrasound in lithotripsy.
• Ionizing radiation—exposure to X-rays and radio-isotopes in X-ray departments, dental clinics, pharmacy and laboratories. Lasers are a form of non-ionizing electromagnetic radiation used in surgery and dermatology, ophthalmology and gynaecology. They can result in eye damage of the user and bystanders from inadequate eye protection and inappropriate direction of the laser beam.

Chemical
• Solvents—xylene in histopathology laboratories, perchloroethylene in hospital laundries and white spirit used by hospital painters are examples of organic solvent exposure in health-care workers.
• Aldehydes—exposure to formaldehyde in post-mortem rooms and glutaraldehyde in endoscopy suites has been associated with respiratory and skin problems. Glutaraldehyde in particular caused concern because of exposure resulting occupational asthma. Environmental hygiene

measurements have, however, shown airborne glutaraldehyde levels below the UK 1991 occupational exposure limit of 0.2 p.p.m. It may be that this standard is inappropriate for an agent like glutaraldehyde, which can cause respiratory and skin sensitization. Reduction of exposure to such aldehydes has included consideration of alternative agents, improvements in local and general ventilation, and personal protective equipment.

• Mercury exposure in dental surgeries and in sphygmomanometer repair has led to elevated mercury in air levels and evidence of increased urinary mercury in exposed individuals. Attention to complete removal of spilt mercury globules is essential to prevent occupational overexposure.

• Cytotoxic drugs and other pharmaceutical agents require adequate procedures for handling in order to prevent unnecessary exposures in pharmacy and nursing staff.

• Anaesthetic gases require operating theatre scavenging systems.

Biological
• Micro-organisms, e.g. tuberculosis, hepatitis B, human immunodeficiency virus, meningococcal meningitis.

Tuberculosis (TB) The concerns over TB relate mainly to:

1   Detecting the health-care worker with active TB at the pre-employment stage, so that there will be no spread of active TB to patients, especially children in paediatric units and in obstetrics wards. This has previously been done by the use of routine chest X-rays of medical staff at the pre-employment screening. However, the pick-up rate for active TB by this routine use of chest X-rays has been poor, and this has therefore been largely discontinued. An alternate system based on the use of pre-employment tuberculin testing (e.g. Heaf test) has been proposed by the British Thoracic Society. This is being implemented by occupational health departments, and the experience has shown that some modifications to the proposals may be needed.

2   Dealing with staff who have been exposed to an active case of TB in the wards, or staff who work in units where contact with TB patients is likely, as in chest clinics and respiratory medicine wards. A surveillance system based on the use of questionnaires asking about haemoptysis, prolonged cough, night sweats and other symptoms suggestive of TB can be used. This is linked to an active health education programme where exposed health-care staff are informed and reminded of the effects and symptoms that should be reported to and investigated by the occupational health department.

8.3   The health of health-care workers

Hepatitis B Health-care staff who are at risk of regular contact with blood and body fluids should be offered hepatitis B immunization. Vulnerable groups include surgeons, nurses and other support staff involved in invasive procedures, laboratory staff, staff of institutions of mental handicap, and front-line ambulance crew. Two vaccines that have been used include a plasma-derived and a genetically engineered yeast-derived vaccine. In the UK the latter has been used extensively for health-care workers. The schedule for immunization requires three 20 microgram (1 ml) doses to be administered intra-muscularly (deltoid). The second dose is given a month after the first and the third at six months. Checking of antibody levels after the full course will allow non-responders to be identified, and a booster dose or a full course of the plasma-derived vaccine can be given. Counselling and advice on extra care with blood and body fluids is needed for those who do not produce any antibody response. A further booster after a full course of vaccine is probably not needed till five years later.

Human immunodeficiency virus (HIV) The number of cases of AIDS (clinical cases or HIV seroconversion) apparently acquired after occupational contact with body fluids is small in comparison with the cases of hepatitis B acquired this way. This is also true of health-care workers transmitting the infection to patients. Safe systems of work when dealing with blood or other body fluids, whether in the laboratory or in the wards, will contribute to reduction of the risk of transmission of occupational blood-borne infections. The concept of 'universal precautions' proposed by the US Centers for Disease Control suggests that all patients should be assumed infectious for HIV and other blood-borne pathogens, and that good work practice and appropriate personal protective equipment should be applied at all times.

Meningococcal meningitis This infection, due to *Neisseria meningitidis* (a Gram-negative diplococcus), can be occupationally acquired through laboratory contact with infective specimens, or from close contact with an infected patient, e.g. via direct mouth-to-mouth resuscitation. Following such contact oral rifampicin can be administered as prophylaxis. However, it is important that female staff on the oral contraceptive pill be advised on the possible reduction of efficacy of the pill by rifampicin.

Mechanical
• Low back pain is a common cause of morbidity in health-care workers. Occupational groups within the health service that are especially vulnerable

are nurses, ambulance crew, theatre staff, radiographers and other groups required to lift and transport patients or equipment, often in difficult environments where lifting aids are not always available or practical. Low back pain is the most common cause of early retirement of ambulance drivers on grounds of ill health . Unfortunately there are no good predictors of candidates likely to suffer from low back pain. The most useful factor to determine at the pre-placement stage seems to be a previous history of low back pain—particularly one that resulted in prolonged periods away from regular work or home activities.

Psychosocial
• Long work hours for junior hospital doctors is a contributory factor to stress in this occupational group. Within the NHS there is now a programme of reduction of such hours to a maximum of 72 work hours a week for junior doctors.
• Stress in ambulance crews has been associated with their need to respond rapidly to requests for assistance by members of the public, doctors or hospital staff. This affects both front-line ambulance crews and those in control rooms. Front-line crews also have to deal with the difficulties of driving rapidly and yet safely through busy traffic, dealing with multiple victims of accidents and disasters, and coping with bereavement. The availability of counselling skills in occupational health departments for such crew members may help to reduce stress.

8.4   Medical audit

Although the term 'audit' is much in vogue, few occupational physicians when challenged know exactly what is meant by the term. Lip-service is paid to the need for audit but few occupational health services have an effective audit system in place.

Medical audit is variously defined but in essence it is a systematic and scientific analysis of the quality and efficiency of medical care. Its purpose is to improve that efficiency and thus provide a more effective service. The audit may be
• internal
• external
Its components are
• setting of objectives (or standards)
• observation of current practice
• comparison of the observed practice with the standard
• reobservation of the practice in the light of changes made.

The audited activity should be
- common
- important in its effect on patients or resources
- easily defined
- compared with agreed and defined standards
- amenable to change

Health care—the usual audited medical activity—can, like a manufacturing process, be broken down into
- structure—resources available, e.g. people, equipment, money
- process—clinical activities, e.g. what is done to the patient
- outcome—the result of clinical intervention.

In many ways, such procedures should be an automatic, intuitive part of any clinical procedure—pre-employment screening, for example. It is the formal nature of the true medical audit which gives real credence to these intuitive clinical reviews.

8.5   Mental health at work

Despite the long list of chemicals capable of causing organic psychosis, mental illness associated with work usually has no such discrete aetiological factor. Furthermore, attitudes to health are changing and, whereas a generation ago, a stiff upper lip might have been considered the appropriate way of coping with a stressful job, nowadays such mental trauma is rightly considered inadmissible.

A specific problem associated with mental illness at work is the occupational physician's role in dealing with the implications such illness may have for other employees as well as for the company. In another context, Lord Moran, in his book about Winston Churchill, recounted the dilemma he faced over being the medical adviser to a Prime Minister who was inexorably sliding into arteriosclerotic senility. When do you say enough is enough and expose the inadequacies of the patient? Unfortunately, there is no easy answer.

Fortunately, however, most occupational physicians are faced with lesser degrees of mental illness, though they are frequently incapable of dealing adequately with it. Archetypally, it is the salesman who, despite early promise, is now failing to meet his sales targets or the manager promoted beyond the level of his real or perceived competence in whom stress has been noted.

Stress is not just another word for anxiety; it also implies a range of behavioural changes. Any one individual may exhibit (or have elicited) a number of the following stress-related symptoms:

- unusual or misguided aggression
- day-dreaming
- paranoia
- disinterestedness
- illogicality
- irrational or impulsive behaviour
- narrow-mindedness
- indecision (or dogmatism)
- procrastination
- inability to relax.

Furthermore, the stressed individual may start drinking excessively, have accidents, marital problems, clashes with 'authority' or increased absenteeism. These features in turn can lead to

- heart disease
- mental illness
- industrial relations problems (including strikes)
- severe or fatal accidents.

It is important to note that there is a relationship between work stimulus and response. Understimulation is as stressful, in a soporific way perhaps, as overstimulation. The problem about *prolonged* overstimulation is that this can lead to collapse in every sense of the word. The perception of premonitory signs should prevent this. It is, therefore, necessary for the physician and the employee to establish what is an acceptable (or necessary) level of stimulation to enable the person to adapt to the work and thereby achieve his or her maximal potential.

Stress is caused by a variety of factors which may be related to

- task
- role
- intrinsic environmental factors at work (overcrowding, undermanning)
- social factors
- physical factors.

In addition, the individual's position in the hierarchy at work will enhance or limit the importance of these factors. For example, managerial persons are perhaps more likely to suffer role-related stress than supervised persons.

On top of all this, the inherent personality trait of the individual will alter his or her response to these stress factors, type A people being more vulnerable than type B.

Type A people are characterized by some (or all) of the following

- competitiveness

- aggression
- impatience
- hyperactivity
- being 'achievers'
- preoccupation with deadlines.

The type B personality is by contrast much less competitive, more 'laid back', more patient but not, paradoxically, less of an achiever.

How should one deal with stress? One thing that should not be advocated is for the subject to 'take things easy', as that is not usually possible. The individual needs to know his/her own character, learn to recognize stressful symptoms as they arise and be on the look-out for what might prove a stressful situation. Beyond that, encouragement to workers to release tension by playing squash, jogging, yoga, etc., is helpful. The art of relaxation is crucial. To deal with stress in others a manager must, in the words of Beric Wright, first 'know his own devil'. The physician is no different.

Helping the individual cope is, however, only one side of the coin. The job itself needs to be scrutinized to see if it can be altered to suit the individual worker more closely.

8.6   Occupational cancer

Volumes have been written on this subject but space allows only the briefest survey here. A short historical account is followed by an annotated list of some of the characteristics of occupational cancer and by some lists indicating the probable and possible human carcinogens of occupational origin.

Historical perspective

The first recognized association between cancer and occupation was made in 1775 by Percival Pott, a surgeon at St Bartholomew's Hospital, London. He noted an increased incidence of scrotal cancer in chimney sweeps and, rejecting the venereal aetiology popular in his day, thought that the tumour was more likely to be due to the soot. Further observations later linked other components of fossil fuel with skin cancer—notably of the scrotum—but it took many long and arduous laboratory experiments before the link was noted experimentally in the first quarter of the twentieth century. The compounds implicated were all polynuclear aromatic hydrocarbons, examples of which include the following:

Benzo(α)pyrine

Dibenz(α, β)anthracene

In 1895, Rehn described bladder tumours in workers in the aniline dye industries. Subsequently, a range of aromatic amines have been noted to be bladder carcinogens. Examples include the following:

2-Naphthylamine

4-Aminobiphenyl

Benzidine

Methylene-bis-*O*-chloraniline

Despite the length of time that has elapsed since the discovery of these carcinogens, workers are still contracting tumours from these or similar agents and will continue to do so. This is partly due to the legacy of a long latent period (sometimes exceeding 40 years) and partly because effective substitutes of such material have not always been available.

Theories of chemical carcinogenesis
Chemicals form the bulk of the occupationally related tumorigens and it is important to distinguish between the classes of chemical carcinogens. In brief, two broad groups are postulated. The *genotoxic* variety poses a clear qualitative hazard to health as they are capable of altering cellular genetic materials and, thus, should theoretically cause cancer after a single exposure. On the other hand, *epigenetic* carcinogens seem to be without direct effect on genetic material and to require high or prolonged exposure for effect. The genotoxic ones probably have no safe threshold; the epigenetic ones might have.

8.6   Occupational cancer

Table 8.1 Classification of chemical carcinogens

Type	Mode of action	Example
Genotoxic		
Direct action	Interacts with DNA	Bischloromethyl ether
Secondary action	Requires conversion to direct type	2-Naphthylamine
Inorganic	Affects DNA replication	Nickel
Epigenetic		
'Solid-state'	Mesenchymal cell effects	Asbestos
Hormone	Endocrine effect ± promoter	Diethylstilboestrol
Immunosuppressor	Stimulates certain tumour growth	Azathioprine
Co-carcinogen	Enhances genotoxic types when given at same time	Ethanol
Promoter	Enhances genotoxic types when given subsequently	Bile acids

A tentative classification which seems reasonable in animal models might thus be expressed as in Table 8.1.

Characteristics of occupational carcinogens
Tumours of occupational origin are usually indistinguishable, histopathologically and symptomatically, from non-occupational tumours. Nevertheless, there are some characteristics of note:
1   They tend to occur earlier than 'spontaneous' tumours of the same site.
2   Exposure to the putative agent is repeated but not necessarily continuous.
3   The latent period is 10–40 years.
4   The tumours are often multiple in a given organ.
5   Despite wildly differing estimates of the proportion of all cancers caused by occupation, the true figure probably lies in the range of 3–8%. However, in the case of asbestos, for example, there is good evidence that life-style and cigarette smoking increase the carcinogenic potential of the asbestos fibre. It is possible that occupation may play a contributory role in apparently spontaneous cancers. Multiple chemical exposures, which are the norm in modern industry, make it, however, exceedingly difficult to isolate particular individual chemicals as the guilty parties and the role of synergism is probably important, though still largely unquantified.

Known or suspected occupational carcinogens
The length of the list and its individual members vary according to which organization has compiled it. Perhaps the most widely accepted list is that

of the International Agency for Research on Cancer (IARC). They regularly review the evidence for carcinogenicity of any compound, where published data exist suggesting a cancer effect. Their evaluations depend on animal data, short-term mutagenicity tests and human evidence, with the greatest weight given to sound epidemiological evidence.

The compound or chemical is then graded for its carcinogenicity into one of four categories

• sufficient
• limited
• inadequate
• lacking

The overall evaluation has five groupings

• *Group 1*

The agent (mixture) is carcinogenic to humans. The exposure circumstance entails exposures that are carcinogenic to humans.

• *Group 2A*

The agent (mixture) is 'probably' carcinogenic to humans. The exposure circumstance entails exposures that are 'probably' carcinogenic to humans.

• *Group 2B*

The agent (mixture) is 'possibly' carcinogenic to humans. The exposure circumstance entails exposures that are 'possibly' carcinogenic to humans.

• *Group 3*

The agent (mixture, exposure circumstance) is not classifiable as to its carcinogenicity to humans.

• *Group 4*

The agent (mixture, exposure circumstance) is probably not carcinogenic to humans.

Table 8.2 lists the current (1990) IARC list of Group 1 carcinogens.

In addition to (or in conjunction with) the list in Table 8.2, there are studies that suggest links between certain tumours and certain industries or materials. Table 8.3 lists the occupations recognized as presenting a carcinogenic risk.

The lists are, however, heavily dependent on the quality of the epidemiological studies.

Undoubtedly, future research will reveal new carcinogens but it is to be hoped that some of the present group will become historical relics. Pre-market testing of new chemicals may help to limit the carcinogenic load for future generations of workers, whilst substitution or stringent environmental control should lessen the impact of known carcinogens. The problem is not an easy one to solve but, whereas tumours related to

8.6   Occupational cancer

Table 8.2 IARC list of Group 1 carcinogens (1990)

Chemicals	Industrial processes	Medicinals	Other
4-Amino biphenyl	Aluminium production	Azothioprine	Aflatoxins
Arsenic and arsenic	Auramine (manufacture)	N, N-Bis	Alcohol
compounds	Coal gasification	(2-chloroethyl)	Betel nut
Asbestos	Coke production	2-naphthylamine	(with tobacco)
Benzene	Haematite mining (U/G)	Chlorambucil	Tobacco
Benzidine	(with radon exposure)	Ciclosporin	
BCME	Iron and steel founding	Cyclophosphamide	
CCME	Isopropyl alcohol	Melphalan	
Chromium	manufacture	Methyl CCNU	
compounds	(strong acid process)	MOPP	
(hexavalent)	Magenta (manufacture)	Myleran	
Coal tar-pitches	Painting	Oestrogens (some	
Coal tar	Rubber industry	steroidal and	
Erionite	Wood (furniture and	non-steroidal)	
Mineral oils	cabinet making)	Oral contraceptive	
(untreated and		(combinations)	
mildly treated)		Phenacetin (in	
Mustard gas		analgesic mixtures)	
2-Naphthylamine		Thiotepa	
Nickel compounds		Treosulfan	
Radon (and its			
decay products)			
Shale oils			
Soots			
Talc (containing			
asbestos)			
Vinyl chloride			

Table 8.3 Occupations recognized as presenting a carcinogenic risk

Industry	Occupation	Site of tumour	Likely carcinogen
Agriculture, forestry, fishing	Farmers, seamen	Skin	Ultraviolet light
	Vineyard workers	Lung, skin	Arsenical insecticides
Mining	Arseniferous ores	Lung, skin	Arsenic
	Iron ore	Lung	? Radon
	Tin	Lung, bone marrow	? Radon
	Asbestos	Lung, pleura and peritoneum	Asbestos
	Uranium	Lung	Radon
Petroleum	Wax pressmen	Scrotum	Polynuclear aromatics
Painting	Painter	Lung, ? bladder	?
Metal	Aluminium production	Bladder, ? lung	Polynuclear aromatics
	Copper smelting	Lung	Arsenic
	Chromate production	Lung	Chromium compounds
	Chromium plating	Lung	Chromium compounds

8.6   Occupational cancer

Industry	Occupation	Site of tumour	Likely carcinogen
	Ferrochromium production	Lung	Chromium compounds
	Iron/steel production/ founding	Lung	Benzo (α)pyrine
	Nickel refining	Nasal sinuses, lung	Nickel compounds
Transport	Shipyards	Lung, pleura and peritoneum	Asbestos
Chemicals	BCME and CMME production	Lung	BCME, CMME
	Vinyl chloride production	Liver	VCM
	Isopropyl alcohol (manufacture by strong acid method)	Paranasal sinuses	?
	Chromate pigment production	Lung	Chromium compounds
	Dye manufacture and users	Bladder	Benzidine, 2-naphthylamine, 4- amino diphenyl, etc.
	Auramine manufacture		
	Poison gas manufacturers	Lung	Mustard gas
Pesticides	Arsenical pesticides	Lung	Arsenic
Gas	Coke plant	Lung	Benzo (α)pyrine
production	Gas retort house	Lung, bladder	Benzo (α)pyrine, 1,2-naphthylamine
Rubber	Rubber manufacture	Lymphatic and haemopoietic systems, bladder	Benzene Aromatic amines
	Calendering, tyre curing and tyre building	Lymphatic and haemopoietic systems	Benzene
	Cable makers, latex producers	Bladder	Aromatic amines
Construction maintenance	Insulators, demolition engineers	Lung, pleura and peritoneum	Asbestos
Leather	Boot and shoe makers	Nose, bone marrow	Leather dust, benzene
Wood pulp and paper	Furniture makers	Nasal sinuses	Wood dust
Electric/ electronics	Engineers	Bone marrow, Brain	?
Health-care industry	Pharmaceutical manufacturers Pharmacists Nurses	Bone marrow	Cytotoxic drugs
	Radiologists/ radiographers	Bone marrow	Ionizing radiation

life-style are difficult to control and many tumours are of unknown
aetiology, all occupationally related cancers are, by definition, theoretically
preventable.

8.7   Repetitive strain injury (RSI)

Synonyms
- Repetitive motion injuries
- Cumulative trauma disorders (CTD)
- Occupational overuse syndrome (OOS)
- Work-related repetitive movement injury (WRRMI)
- Work-related upper limb disorders.

Definition
The definition of the condition is controversial and confused. RSI usually
refers to pain and discomfort in the upper limbs as a result of repetitive
movements or constrained postures. The term RSI has encompassed a
whole variety of clinical entities, primarily in the wrists and elbows, which
have arisen mainly from work activities. These include
- tenosynovitis
- de Quervain's disease
- carpal tunnel syndrome
- peritendinitis crepitans
- epicondylitis.

Some have restricted it to effects in the upper limbs only. Others refer
to RSI also affecting the neck, shoulders, and even the lower limbs. Case
definitions used may allow for both occupational and non-occupational
activities, or may be confined to situations where only occupational factors
are involved.

Main clinical features
- Persistent pain and discomfort of the muscles, tendons or soft tissues of
the limbs.
- Swelling, tenderness, inflammation and crepitus on examination.
- Tingling and numbness, especially with nerve involvement.
    Symptoms may be acute, recurrent or chronic.

8.7   Repetitive strain injury (RSI)

Relevant exposures
• Dynamic load—repeated, frequent movements of the limbs.
• Movement through the full range of motions, e.g. pronation to supination and back.
• Awkward postures and/or postures at the extremes of the range of movements, e.g. hyperflexion or hyperextension, or ulnar/medical deviation at the wrist.
• Static load—e.g. holding the limbs against force or pressure for sustained periods of time.

Other contributory factors
Many of these factors are debatable, especially the extent to which they contribute to the prevalence of RSI.
• previous injury?
• individual susceptibility—some cases have occurred in part-time rather than full-time workers, thereby suggesting that work activity is not the only determining factor in whether RSI develops
• psychological overlay—especially with monotonous work, or bonus rate dependent on output, or work performance under peer pressure from team members
• perceived ease of obtaining monetary benefits under compensation schemes available.

Extent of the problem
Tenosynovitis is the second most common prescribed disease in the UK (after occupational dermatitis).
    Some occupations documented as having a high prevalence of RSI are
• poultry processing
• electronics assembly
• mechanical components assembly line
• data processing
• telephonists
• checkout operators at supermarkets and stores.
    It appears to be more common in women and blue collar workers.

Treatment
• Surgery.
• Rest with immobilization of the affected area.
• Physiotherapy.
• Analgesics and NSAIDs (non-steroidal anti-inflammatory drugs).

• Ultrasound.
• Infra-red treatment.
• Local injection with hydrocortisone.
    These procedures can only be effective if accompanied by improvements in the job design or in the ergonomics at the workstations.

Prevention
• Attention to ergonomics at workplace.
• Proper design of machinery, tools and equipment.
• Adequate training of operators.
• Rest periods—views on this range from allowing rest breaks of 5 minutes every 1 to 2 hours, to not having a rigid system of enforced rest breaks but concentrating on flexibility in job design thereby allowing employees to take rest breaks as needed.
• Job rotation with work involving different physical effort and activities.

8.8   Rehabilitation and resettlement
The rehabilitation and resettlement of disabled employees is an integral part of the work of an occupational health service. This process may involve
• medical and surgical measures
• physiotherapy
• retraining
• advice on social and family readjustment
• relocation in new or previous employment.
    To be of maximum use to the employee seeking re-employment, the occupational health team must be familiar with
• the social services at their disposal in the community
• the benefits (financial and otherwise) to which the employee is entitled
• the nature of the job, its skills, safety, hazard, strenuousness.
    All this sounds straightforward, except that, in the UK at least, the services are disparate, incomplete and bewilderingly complex. Furthermore, the agencies change names and roles with disturbing alacrity.
    Some definitions are, however, clear-cut:
• *impairment*—lacking part (or all) of a limb or mechanism
• *disability*—functional incapacity or limitation
• *handicap*—disadvantage or restriction of activity.
These can, obviously, be related to each other.
    The problem of disablement is not a small one. It is estimated that currently around 5–12% of the adult populations of most Western

countries have some degree of 'limitation', with about half that number having a severe handicap. Between the ages of 50 and 64, over a quarter of those handicapped suffer from arthritis in one form or another and a further quarter have cerebrovascular or neurological disease. This age-group constitutes a quarter of the total adult population with severe disablement.

In the UK, the Government involvement in employee rehabilitation and resettlement is through the Department of Employment, who operate the Job Centres and Employment Offices. The unemployed disabled person is resettled through the Disablement Resettlement Officer (DRO), who is based at the Job Centre. The DRO's main function are
- registration of the disabled
- job finding
- adaptation of workplaces
- the provision of special aids to employment
- referral to an Employment Rehabilitation Centre (ERC) for training.

The Health and Safety Executive Medical Division has an advisory role to the disablement services, largely to assist this process of rehabilitation and resettlement.

9 Control of airborne contaminants

9.1   Introduction, 239

9.2   Extract ventilation design, 244

9.3   Dilution ventilation, 256

9.4   Choice of fan, 260

9.5   Air cleaning and discharges to the atmospheres, 267

9.6   Energy and cost implications of ventilation systems, 268

9.1   Introduction

The purpose of applying workplace control techniques is to minimize worker exposure to the potential hazard, ideally so that exposure levels are below what is considered to be hazardous. The success of the selected control will be judged by its own ability to reduce personal risk. The control method should remain effective and maintain the same degree of protection over the working life of the process. Potential risk should be assessed by examining the results of failure of the control system. Where chronic hazards exist, an occasional transient failure may not be too serious but any overexposure could have serious implications in the cse of sensitizers and carcinogens, and, in the case of asphyxiants, could be fatal. Thus, the control system must be designed to match the potential risk and, when risks are great, tolerances and safety factors should be designed accordingly. Where potential failure could have serious consequences, there will be a need to build in redundancy and back-up controls to warn of overexposure.

In accordance with good occupational hygiene practice, the control systems will take two forms—software and hardware—which can be applied together or separately.

Software consists of
- substitution of a less hazardous material
- methods of work to reduce worker exposure
- training of workers to adopt safer methods of work
- application of work schedules or regimes to limit exposure times.

Hardware consists of
- enclosure of the process
- suppression of emissions
- shielding of source or worker
- ventilation, extract at source
- ventilation, dilution to reduce concentration
- shielding of the worker from the emission
- application of personal protective clothing to the worker.

As far as exposure to substances is concerned, there is a legal duty to prevent or control enshrined in Regulation 7 of the COSHH Regulations (see p. 33). Paragraph 34 in the Approved Code of Practice sets out the order in which the procedure should be tackled as follows:

For preventing exposure (software solutions):

1   Eliminate the use of the substance.

2   Substitute by a less hazardous substance or by the same substance in a less hazardous form.

9   Control of airborne contaminants

9.1   Introduction

For controlling exposure (hardware solutions):

1   Totally enclose process and handling systems.

2   Design the plant, process or systems of work to minimize the generation of the substance or suppress or contain it.

3   If spills are likely then design the area to minimize the area of contamination.

4   Partially enclose the source with local exhaust ventilation.

5   Provide local exhaust ventilation.

6   Provide sufficient general ventilation.

For controlling exposure (software solutions):

1   Reduce the number of employees present.

2   Exclude non-essential access.

3   Reduce the period of exposure for employees.

4   Regularly clean contamination from, or disinfect, walls, surfaces, etc.

5   Provide safe means of storage and disposal of substances hazardous to health.

6   Provide suitable personal protective equipment.

7   Prohibit eating, drinking, smoking, etc. in contaminated areas.

8   Provide adequate facilities for washing, changing and storage of clothing, including arrangements for laundering contaminated clothing.

When tackling a workplace health hazard with a view to reducing the risk, using engineering controls, the order in which the problem should be approached is as follows:

1   Deal with the *source* of emission or hazard.

2   Examine the *transmission* of the hazard between the source and the worker.

3   Protect the *worker* or the exposed *population*.

The hazard can be

• an airborne pollutant, such as dust, gases, vapours, a microbiological organism or a radioactive particle, all of which enter the body via the lungs

• a radiated emission, such as noise, heat, light, ionizing and non-ionizing radiation, which affects the body through the skin or other exposed organs

• a chemical in liquid or solid form, which can affect the skin or enter the body via that organ.

Source

The source can be tackled, in the case of airborne pollutants, to reduce the potential for emission, as follows:

• substitution of the toxic material for one with a lower hazard potential

9.1   Introduction

• changing the process so that no hazard is created
• enclosing the point of emission to minimize the area of outlet openings
• providing extraction ventilation to capture the material at the point of release
• suppression at source by wet methods or quenching techniques.
    The emission of radiation can be approached as follows:
• reduce the intensity of the source
• change the wavelength of radiation to a safer one
• enclose the point of emission
• attenuate at the point of emission.

Transmission

With airborne pollutants, once the material is airborne and away from the source, control involves
• shielding between the worker and the source
• application of dilution ventilation
• the use of jet ventilation to divert the contaminated air.
    In the case of radiated emissions:
• increase the distance between the worker and the source
• attenuate the radiation
• deflect or divert radiation.

Exposed population

The workers' exposure can be minimized by examining their position in relation to the hazards.
    In the case of air pollution:
• enclose the worker
• eliminate the need for a worker, using remote control of the process or automation
• provide respiratory protection
• wash the worker in a stream of uncontaminated conditioned air
• reduce the duration of exposure by means of job rotation
• apply a safer method of work involving less contact with the pollutant
• educate and train the worker to appreciate the hazards so that his/her own behaviour will minimize the exposure.
    With radiation and skin contact:
• enclose the worker in a protective cabin or behind shields
• remove the worker by means of remote control operation of the process or by automation
• provide protective clothing to all vulnerable or exposed parts of the body
• apply a safer method of work

• educate and train the worker in the risks involved and the application of safer methods of work.

Sources of emission

Sources of emission can be either *predictable* in time and space, being continuous or periodic, or *unpredictable* (fugitive), occurring haphazardly due to breakdown or wear of normal engineering items. The former can be dealt with by engineering methods, some of which are outlined here, but the latter cannot easily be dealt with and normally involve the use of personal protective equipment.

Periodic or continuous sources of emission occur from the following operations:

1   Material handling which gives rise to dust, gases or vapours, for example:
    debagging
    pouring of liquids
    transfer from one container to another
    transfer from one mode of transport to another
    blending
    stirring or agitating
    screening or sieving
    crushing or grinding
    emptying
    recharging
    sampling

2   Processes causing emission of (mainly) vapours but also particles:
    stirring and agitating
    surface coating
    drying
    spraying
    dipping
    curing
    baking
    welding
    sampling

3   Processes producing mainly dust but also some fume or gas:
    machining
    drilling
    planing
    sanding
    milling

9 Control of airborne contaminants

9.1 Introduction

cutting
sawing
dismantling
demolition.

Fugitive emissions produce gases, liquids and vapours and occur as a result of leaks from

- fractured or corroded pipes, vessels and containers
- poorly made joints or the breakdown of seals on joints
- along the shafts of valves
- along the shafts of pumps
- spills and collisions.

Maintenance of plant and equipment always carries a great risk of predictable and unpredictable exposure and maintenance workers require special attention and equipment to minimize their exposure. Software should be particularly attended to, because maintenance workers work unsupervised in places where access is difficult and at times when safety and health staff are not on site.

Control of periodic and continuous emissions

It is necessary to examine the working process carefully to establish where the sources of emission are and how an improvement can be effected. In most cases, the emission of dust and vapours can be minimized by enclosure or by redesigning the process so that escape of pollutants is reduced. The openings where pollutants escape can be fitted with doors or covers which remain closed most of the time but which are open for access. If this is impracticable, then the openings for access can be fitted with extract ventilation but it is worth remembering that the smaller the openings, the lower the air-flow rate required to capture the pollutants and, hence the cheaper the costs.

Redesign of the process can be simple, for example:

- fitting covers on containers and openings
- matching discharge ports to entry holes
- using sealed transfer systems
- using anti-splash discharge nozzles.

A more expensive solution is to completely enclose and automate the process, thus keeping the bulk of the contaminants inside, although eventually in most processes the end-product has to be exposed to atmosphere as do the raw materials at the start. With highly toxic materials, automation may be the only solution.

The provision of enclosure means that visibility of the process is impaired and, where it is necessary to observe the process, enclosure

materials should be transparent. Closed-circuit video viewers could be adopted. Observation of the process is often necessary to judge whether a container is full or empty; thus visibility is not important if an alternative means of indication is adopted. This may involve one of the following:
• placing the vessel or container on a weight indicator
• using a float indicator
• using a beam of light or a stream of radioactive particles to a sensor placed on the opposite side of the vessel. The intervention of the material cuts the beam, thus indicating that it has reached that level
• using pressure switches.

Process redesign is often a cheaper expedient than the provision of extract ventilation, which requires skilful design to be successful and is costly to build and to operate, particularly with regard to replacing heat removed by the discharge of extracted air to atmosphere. The costs of heating 'make-up' air are given later in this chapter.

9.2   Extract ventilation design

Where alternative solutions cannot be adopted, the pollutants can be captured before they are generally released into the working environment by means of extract ventilation. This can be achieved by some kind of hood, enclosure or slot, sufficiently negatively pressurized to ensure an inward current of air that will carry with it the airborne pollutants. The extract device will normally be connected to a fan via ducting and thence to a point of discharge. An air-cleaning system in the duct may be necessary to ensure that the discharged air is sufficiently clean, either to recirculate to the workplace or to satisfy external environmental standards, if released outside.

With an unrestricted suction inlet such as a hood, air will flow from all sides, in a zone of influence which is approximately spherical, with the inlet at the centre. Thus, with unflanged hoods, air will flow in from behind the inlet as well as in front where there may not be any pollutants and so this air is wasted. Flanges and screens are, therefore, necessary to channel the air at a sufficient velocity over the point of release of the pollutant, to ensure successful capture.

Capture velocity is defined as that velocity which will overcome the motion of the airborne pollutant to draw it into the mouth of the extract. The following factors will influence the capture velocity
• the velocity of release of the pollutant
• the degree of turbulence of the air around the source
• in the case of particulates, their aerodynamic diameter
• the density of the materials released.

9.2   Extract ventilation design

Table 9.1 Recommended capture velocities

Source conditions	Typical situations	Capture velocity (m s^{-1})
Released into still air with no velocity	Degreasing tanks, paint dipping, still air drying	0.25–0.5
Released at a low velocity or into a slow moving airstream	Container filling, spray booths, screening and sieving, plating, pickling, low speed conveyor transfer points, debagging	0.5–1.0
Released at a moderate velocity or into turbulent air	Paint spraying, normal conveyor transfer points, crushing, barrel filling	1.0–2.5
Released at a high velocity or into a very turbulent airstream	Grinding, fettling, tumbling, abrasive blasting	2.5–10.0

Recommended capture velocities are given in Table 9.1.

Capture distance is the distance between the point of release and the mouth of the inlet.

Face velocity is the air velocity across the mouth or face of the inlet.

Aspect ratio is the ratio of the width of the face divided by the length.

With booths, enclosures and fume cupboards, face velocity is used rather than capture velocity. Sufficient face velocity should be provided to prevent the pollutants released inside the enclosure from escaping back towards the worker. Recommended face velocities will depend upon the degree of toxicity of the material being handled and the amount of air turbulence found at the entrance and will vary from 0.2 m s^{-1}, for an aerodynamically shaped inlet handling low toxicity materials, up to 1.5 m s^{-1} for the opposite. It must be borne in mind, however, that with non-aerodynamically shaped entrances, the higher the air velocity, the greater the degree of turbulence created by the inlet.

The shape of the extraction inlet will be dictated by the shape of the workplace and the area over which the pollutants are released. If the suction inlet area is too large, air distribution across the face may be uneven, allowing pollutants to escape capture over certain parts of the emission area. If the larger face dimension exceeds 1.5 m it is advisable to provide twin duct offtakes or split the area with two hoods. It is advisable for flow splitters to be fitted to any hood where velocity distribution is expected to be uneven. Table 9.2 shows the shapes and uses of different suction inlets.

9.2   Extract ventilation design

Table 9.2 Suction inlet shapes and uses

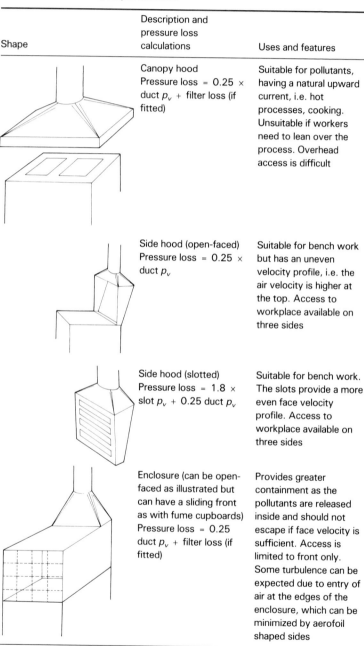

Shape	Description and pressure loss calculations	Uses and features
	Canopy hood Pressure loss = 0.25 × duct p_v + filter loss (if fitted)	Suitable for pollutants, having a natural upward current, i.e. hot processes, cooking. Unsuitable if workers need to lean over the process. Overhead access is difficult
	Side hood (open-faced) Pressure loss = 0.25 × duct p_v	Suitable for bench work but has an uneven velocity profile, i.e. the air velocity is higher at the top. Access to workplace available on three sides
	Side hood (slotted) Pressure loss = 1.8 × slot p_v + 0.25 duct p_v	Suitable for bench work. The slots provide a more even face velocity profile. Access to workplace available on three sides
	Enclosure (can be open-faced as illustrated but can have a sliding front as with fume cupboards) Pressure loss = 0.25 duct p_v + filter loss (if fitted)	Provides greater containment as the pollutants are released inside and should not escape if face velocity is sufficient. Access is limited to front only. Some turbulence can be expected due to entry of air at the edges of the enclosure, which can be minimized by aerofoil shaped sides

9.2   Extract ventilation design

Shape	Description and pressure loss calculations	Uses and features
	Booth with top extract and bottom supply through a perforated or gridded work surface Extract pressure loss = 0.25 × duct p_v	Provides good capture of internally produced pollutants and ensures low entry velocities through opening, thus minimizing turbulence. Careful design is required to balance air flow rates to ensure a 10–25% excess of extract over supply. This ensures a small inflow at the mouth
	Slot, defined as having an aspect ratio of more than 5:1 Pressure loss = 1.8 × slot p_v + 0.25 × duct p_v	Has good access all round, suitable for surfaces of tanks or where pollutant release is spread over an area rather than from a point. Capture distance is limited
	Double slot with extraction from two sides Pressure loss = 1.8 × slot p_v + 0.25 × duct p_v	Has good access all round, suitable for surfaces of tanks or where the pollutant is spread over a wide area. A double slot will always provide better capture than a single one
	Extract hood with supply slot Extract pressure loss = 1 × face p_v + 0.25 × duct p_v	Suitable for control over very wide surfaces as the supply of air can sweep the pollutants into the extract. Careful design is required to balance air-flow rates to ensure that more air is captured than supplied. Also the 10° expansion of a jet of air must be borne in mind when sizing the extract hood

9.2   Extract ventilation design

Shape	Description and pressure loss calculations	Uses and features
	Portable hood on flexible duct Pressure loss = 0.25 × duct p_v	Suitable for sources of pollution that are moving, such as welding on a large workpiece
	Curved slot Pressure loss = 1.8 × slot p_v + 0.25 × duct p_v	Suitable for extracting closely around containers
	Extract annulus Pressure loss = 1.8 × annular p_v + 0.25 × duct p_v	Suitable for extracting around discharge pipes and outlets
	Evacuated containers Pressure loss = 0.65 × duct p_v	Suitable for sources of emission which are released at very high velocities. An evacuated container is placed in the path of the particles as they are emitted. Capture velocities about 10 m s^{-1} are required

p_v refers to velocity pressure at the stated position.
Note: The performance of many of the above suction inlets can be improved by the addition of a flange fitted around the periphery of the inlet. This has the effect of limiting the zone of influence to the immediate area in front of the inlet. The effect this has on capture velocity and distance is shown on p. 249.

Low-volume, high-velocity extract systems

This technique is useful for extraction on portable hand-held power tools, such as grinders, circular saws and sanders. The principle is to place the suction inlet as close as possible to the point of release of the pollutants. Given a capture velocity of around 15–30 m s^{-1}, the closer the inlet to the point of release, the less volume flow required and the smaller the ductwork. For hand-held tools, the ducting must be light and flexible and is made of plastic, ribbed with reinforcing material and no larger in diameter than the hose of a domestic vacuum cleaner. The collected pollutants are usually particles of metal or wood, which are collected in an industrial type of vacuum cleaner. The whole unit is, thus, portable and self-contained.

9.2   Extract ventilation design

Suction inlet performance

With straight-sided hoods and slots, centre-line velocities in the area in front of the inlet can be predicted. They are dependent upon the aspect ratio of the inlet, the mean face velocity and the distance from the hood. The relationship developed by Fletcher is given in the expression:

$$\frac{V}{V_0} = \frac{1}{0.93 + 8.58\,\alpha^2}$$

where $\alpha = XA^{-1/2}\left(\dfrac{W}{L}\right)^{-\beta}$

$\beta = 0.2(XA^{-1/2})^{-1/3}$

V = centre-line velocity at distance X from the hood mouth

V_0 = mean velocity at the face of the hood

L = length of hood

W = width of hood

$A = LW$.

The volume flow rate through the hood can be calculated by assigning V to the required capture velocity and X to the designed capture distance. The volume flow Q is then obtained from:

$$Q = V_0 A$$

This formula also applies to the relationship between volume flow rate, cross-sectional area and air velocity in ducts. The solution of the Fletcher equations can be simplified by using the nomogram provided in Fig. 9.1.

The addition of flanges to the hoods will increase centre-line velocities by up to 25% for hoods with a low aspect ratio, but up to 55% with a slot of aspect ratio 16 : 1. Optimum flange widths are in the region of $A^{-1/2}$.

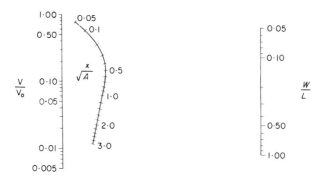

Fig. 9.1 Nomogram for the solution of the Fletcher equations (Crown copyright).

If capture distances are too great, the required face velocities to provide a suitable capture velocity can become excessive. For example, to provide a capture velocity of 1 m s^{-1} at a distance of 1 m from the mouth of a 200×1000 mm slot would require a face velocity of 66 m s^{-1}, which is unrealistic.

Prediction of velocities away from the centre-line is difficult. In general, however, they will be less than the centre-line velocity on either side.

Unfortunately, the theoretical prediction of centre-line velocities assumes that the surrounding areas are free from disturbing air currents, which is often far from the case, particularly if there is a regular movement of people and vehicles in the vicinity of the source of pollution. Also, the edges of the hood, the workpiece and the worker all contribute to local air turbulence, which can result in pollutants being drawn out of the capturing airstream and into the workplace and the breathing zone of the worker. In an ideal workplace, the entries to hoods should be of aerofoil shape to minimize turbulence. In practice, one of the few situations where aerofoil shapes are used is in modern fume cupboards, where they have been shown to be successful in minimizing the escape of pollutants. It is possible to add aerofoils to the entrances to older fume cupboards and they have been shown to improve capture.

Ducts and fittings

Air is conveyed from the extract inlet to the point of discharge in ducts, which take a route to suit the needs of the building in which they are housed. In so doing, fittings such as bends and changes of section are required to assist in negotiating obstructions and to accommodate items of equipment, such as filters and air cleaners. The remaining parts of the system employ straight ducts. The shapes and sizes of the ducts depend upon the configuration of the workplace and the building and the desired velocity of the air inside the ductwork. There are several factors that should be considered when deciding upon the cross-section of the ducts and, usually, a compromise is made between them. For a given volume flow rate, the larger the duct, the lower the air velocity inside and the less energy absorbed in overcoming friction. Such a duct is, however, high in capital cost. A circular cross-section is more economical in material than a rectangular one but, in some buildings, the space available into which the duct can be placed is more suited to the rectangular shape.

If dust or larger particles are to be conveyed, then it is important to ensure that it does not settle out and deposit inside the ducts to cause an obstruction or, in the case of flammable dusts, become a potential

9.2   Extract ventilation design

Table 9.3 Recommended transport velocities

Pollutant	Transport velocity ($m\ s^{-1}$)
Fumes, such as zinc and aluminium	7–10
Fine dust, such as lint, cotton fly, flour, fine powders	10–12.5
Dusts and powders with low moisture contents, such as cotton dust, jute lint, fine wood shavings, fine rubber dust, plastic dust	12.5–17.5
Normal industrial dust, such as sawdust, grinding dust, food powders, rock dusts, asbestos fibres, silica flour, pottery clay dust, brick and cement dust	17.5–20
Heavy and moist dust, such as lead chippings, moist cement, quick-lime dust, paint spray particles	over 22.5

explosion hazard. The correct transport velocity is, therefore, required to minimize deposition.

Transport velocity is the minimum air velocity within the duct necessary to maintain the particles airborne. Table 9.3 gives recommended transport velocities.

If only gases and vapours are to be carried, then transport velocities are not important and the air velocity becomes a matter of economics or acoustics. Optimum velocities are usually between 5 and 6 $m\ s^{-1}$ but, if noise levels are not to be obstrusive, 5 $m\ s^{-1}$ should be the maximum.

In extract ventilation design, the volume flow rate (Q) is determined first by virtue of the requirements of the suction inlet. Then, having decided upon the duct velocity (V), the cross-sectional area (A) is determined from the expression

$$Q = VA$$

Ducting is usually made of galvanized sheet steel but a variety of other materials can be used, including brick, concrete, PVC, fibreglass, canvas, plastic, stainless steel. Where the air contains corrosive materials, galvanized metal is unsuitable and a corrosion-resistant material must be used.

Pressure losses

Air requires a pressure difference for it to flow and it will always flow from the higher to the lower pressure. The source of motive power is either natural or by means of a fan. Pressure is a type of energy which appears in two forms: static (p_s) and velocity (p_v), and the sum of these is known as

9   Control of airborne contaminants

9.2   Extract ventilation design

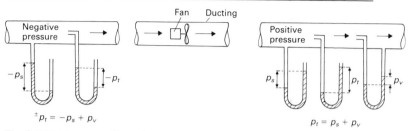

Fan   Ducting

Negative pressure

Positive pressure

$$^{\pm}p_t = -p_s + p_v$$

$$p_t = p_s + p_v$$

Fig. 9.2 U-tube gauges illustrating pressures around a fan (after Waldron and Harrington (1980) *Occupational Hygiene*, Blackwell Scientific Publications, Oxford, p. 115).

total pressure (p_t). Static pressure is exerted in all directions by a fluid that is stationary but, if it is moving, it is measured at right angles to the direction of flow, to eliminate the effects of velocity. Static pressure can be either positive or negative in relation to atmospheric pressure: on the suction side of a fan it is usually negative and on the delivery side it is normally positive. This is illustrated using the U-tube gauges shown in Fig. 9.2.

Velocity pressure is the kinetic energy of a fluid in motion and is calculated from the following expression:

$$p_v = \frac{\rho v^2}{2}$$

where ρ is air density and v is air velocity.

If standard air density at 1.2 kg m^{-3} is used, then the above expression becomes

$$p_v = 0.6v^2$$

If v is in metres per second, p_v will be in Pa. These expressions are widely used in air-flow measurement and in the calculation of pressure losses in ductwork systems.

Reynolds number

In a pipe through which fluid flows, the relationship between the energy absorbed and the rate of flow depends upon the character of the flow. At low velocities, a non-turbulent flow exists which is termed 'laminar' or 'streamlined' flow, where the energy absorbed is proportional to the velocity of the fluid. At higher velocities, a turbulent flow exists in which the energy absorbed is proportional to the square of the fluid velocity. The character of the flow is determined by the following variables:

μ = the dynamic vicosity of the fluid
ρ = density of the fluid
v = velocity of the fluid
D = diameter of the pipe.

These variables combine together to form a dimensionless number known as Reynolds number (R_E) which may be calculated from the equation

$$R_E = \frac{vD\rho}{\mu}$$

In general with $R_E < 2000$, lamninar flow exists and, with $R_E > 4000$, turbulent flow exists. Between these limits, flow conditions are variable. In most engineering applications, $R_E > 4000$ and it is assumed that the energy absorbed is proportional to the square of the velocity. The exception to this general statement is in air filtration, where air velocities can be very low within the filter medium and R_E approaches 2000.

The energy losses due to friction are expressed as a pressure loss, which can be calculated or obtained from charts and nomograms. With regard to straight lengths of ducting, pressure loss can be obtained in Pa per metre, using the nomogram given in Fig. 9.3. As the duct sides are parallel, there is no change in air velocity from one end to another. The pressure losses obtained from the nomogram are, therefore, both total and static. The losses in most fittings are calculated by multiplying the velocity pressure at a point in the fitting, by a factor determining empirically for the geometric shape of that fitting. The resulting pressure loss is in total pressure. It is important to work in total pressure for ventilation calculations because fittings and changes of section have changes in static pressure within them, sometimes resulting in a gain of static, but a loss of total, pressure. Working in total pressure throughout avoids any confusion.

Fan required

In order to establish the duty of a fan to draw air through the system, it is necessary to sum the individual pressure losses from each of the components, starting from the suction inlet and working towards the fan (see Fig. 9.4). If there are components on the delivery side of the fan, then it is also necessary to add the pressure losses on the side. Similarly, if a filter or dust collector is installed, the pressure loss of that component must be included. This will be obtainable from the manufacturer and should be quoted both for a clean unit and for the time when the filter needs changing or cleaning, the latter higher pressure being included in the

9.2 Extract ventilation design

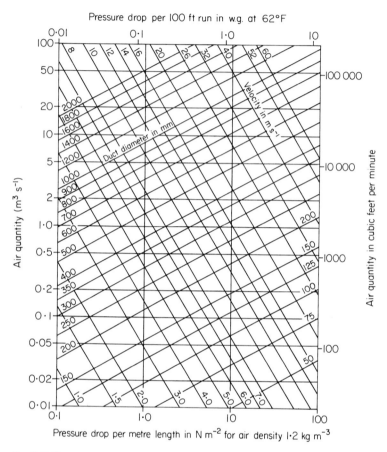

Pressure drop per 100 ft run in w.g. at 62°F

Air quantity (m³ s⁻¹)

Air quantity in cubic feet per minute

Pressure drop per metre length in N m⁻² for air density 1·2 kg m⁻³

Fig. 9.3 Nomogram for calculating air flow in round ducts (plotted on log scales) taken from CIBS Guide Section C4, published by Chartered Institute of Building Services.

calculation. Having summed all the pressure losses involved throughout the system (this should include the discharge velocity pressure), the fan can be chosen from manufacturers' catalogues to handle the chosen volume flow rate at the total pressure calculated.

It is not unusual with extract systems to add several branches, all feeding to a single duct and fan. When this occurs, the pressure required to specify the fan is normally taken as that required to bring the air from the inlet furthest from the fan, through the system to the discharge point. Sufficient pressure will be available to overcome the resistance of the

9 Control of airborne contaminants

9.2 Extract ventilation design

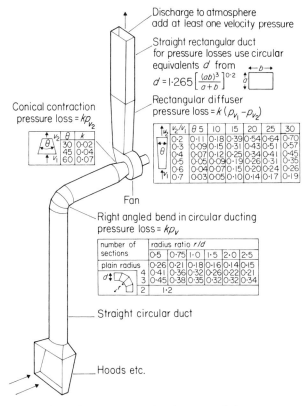

Discharge to atmosphere
add at least one velocity pressure

Straight rectangular duct
for pressure losses use circular
equivalents d from

$$d = 1.265 \left[\frac{(ab)^3}{a+b} \right]^{0.2}$$

Rectangular diffuser
pressure loss $= k\,(p_{v_1} - p_{v_2})$

v_2/v_1 \ θ	5	10	15	20	25	30
0·2	0·11	0·18	0·39	0·54	0·64	0·70
0·3	0·09	0·15	0·31	0·43	0·51	0·57
0·4	0·07	0·12	0·25	0·34	0·41	0·45
0·5	0·05	0·09	0·19	0·26	0·31	0·35
0·6	0·04	0·07	0·15	0·20	0·24	0·26
0·7	0·03	0·05	0·10	0·14	0·17	0·19

Conical contraction
pressure loss $= k p_{v_2}$

θ	k
30	0·02
45	0·04
60	0·07

Fan

Right angled bend in circular ducting
pressure loss $= k p_v$

number of sections	radius ratio r/d					
	0·5	0·75	1·0	1·5	2·0	2·5
plain radius	0·26	0·21	0·18	0·16	0·14	0·15
4	0·41	0·36	0·32	0·26	0·22	0·21
3	0·45	0·38	0·35	0·32	0·32	0·34
2	1·2					

Straight circular duct

Hoods etc.

Fig. 9.4 Pressure losses in components of an extract ventilation system. p_v refers to velocity pressure, the suffix referring to the position as indicated in the diagram accompanying each fitting. For fan, see p. 260. For pressure losses, see Fig. 9.3. For hoods, etc., see pp. 244–248.

intermediate branches, i.e. those nearer the fan. In fact there may be an excess of pressure, making it necessary to restrict the intermediate branches in order to prevent an excess of air flow from passing through them to the detriment of the furthest branch. This results in the multi-branched system being out of balance, a common fault with many industrial systems which have been in use for some time. Balancing can be achieved by installing adjustable dampers in the intermediate branches or by making those branches higher in resistance by design. The damper method has the disadvantage of being thrown out of balance by injudicious tampering unless the damper handles are locked in the balance position. Also, if dust is to be carried in the duct, the dampers can act as a depository, such that unwanted accumulations of dust can build up, making the damper unworkable and causing an unnecessary obstruction in the duct. Inherent balancing by design is, therefore, to be preferred.

When designing a system, it is useful to draw a sketch of the layout, labelling each junction where the air changes speed or direction or where two airstreams meet, by assigning the change point with a number or letter of the alphabet. In this way, each section can be identified as, for example, 3–4 or E–F. A table should be drawn up with headings as in Table 9.4.

The table should be filled in section by section, starting from the inlet furthest from the fan, using the design information selected as appropriate to the requirements of each section. It is not necessary to complete every column for each section, as they do not all apply and it may be necessary to leave some sections until after the following section has been designed. The last column on the right of the table is the sum of the pressure losses accumulated from the beginning and this provides a statement of the pressure inside the duct at that point, in relation to the atmosphere. Thus, if a connection was to be made to the atmosphere at that point, the pressure difference available to overcome the resistance of the connecting branch is known.

System resistance characteristic

At the Reynolds numbers found in most components of an extract system, the pressure loss is proportional to the square of the volume flow rate. Thus, once the pressure loss is calculated for a particular volume flow rate, losses for all other flows can be simply calculated. It follows that the system can be represented by a curve plotted on linear ordinates of the form $y \propto x^2$ (i.e. a parabola), passing through the origin and the calculated duty point (Fig. 9.5). It is normal to plot pressure on the y-axis and volume flow rate on the x-axis and, from this curve, the pressure loss for any other flow can be read. This curve is known as the system resistance characteristic.

9.3   Dilution ventilation

Failure to capture pollutants at source will result in the general workroom air becoming contaminated, exposing a wider population. For worker contact to be minimized, separation or dilution can be adopted. Separation involves the screening or segregation of the workers from the pollutant, which will be invading the workroom as a cloud of gas or particles. This be achieved either by means of hardware or by software techniques,

Table 9.4 Headings for table to be used in designing an extract system

Section	Length (m)	Volume flow ($m^3 s^{-1}$)	Duct dimension (mm)	Duct area (m^2)	Air velocity ($m s^{-1}$)	Velocity pressure (Pa)	Loss factor (k)	Pressure loss per metre ($Pa m^{-1}$)	Section pressure loss (Pa)	Cumulative pressure loss (Pa)

9.3   Dilution ventilation

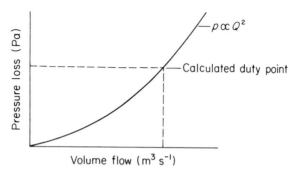

Fig. 9.5 System resistance characteristic.

such as work schedules or production routines, which can be planned to reduce the potential for exposure. More commonly, general ventilation is provided to supply uncontaminated air in sufficient quantities, to dilute the pollutants to what is considered to be a safe level.

Dilution ventilation should be provided only where
* the pollutant is of low toxicity
* only small quantities of pollutant are emitted
* the emission is at a steady and predictable rate
* the ventilation air can reach the airborne material to dilute it before reaching the workers.

Required volume flows Q can be calculated in the following stages:

1   Establish the rate R of the emission of the contaminant, which may involve determining the rate of use of the source material, e.g. paint or solvents, which may evaporate into the workroom over the course of the shift. The amount released will bear a direct relationship to the quantity used.

2   Decide upon the airborne concentration that is acceptable in the workroom. Where a level is recommended by a standard setting authority, such as an OES, some value below that should be used to safeguard against miscalculation or to protect the atypical worker.

3   Determine the dilution factor k that is likely to be required, taking into account the flow patterns of the diluting air, the behaviour of the clouds of polluting material as it is released and the position of the worker in relation to both. The estimation of this figure is a matter of skill, judgement and experience.

The required diluting volume flow of air is calculated from the following:

9   Control of airborne contaminants

9.3   Dilution ventilation

---

1   If C is in p.p.m. and R in $m^3 \, s^{-1}$,

$$Q = \frac{R. \, 10^6.k}{C} \; m^3 \, s^{-1}$$

2   If C is in $mg \, m^{-3}$ and R in $m^3 \, s^{-1}$,

$$Q = \frac{R.D. \, 10^6.k}{C} \; m^3 \, s^{-1}$$

where D is the vapour or gas density of the pollutant in $kg \, m^{-3}$. Note that the liquid density is not the same as the vapour density, which at standard temperature and pressure can be calculated from

$$D = \frac{M}{22.4} \; kg \, m^{-3}$$

where M is the molecular weight of the pollutant.

Summary of the aims of both dilution and extraction ventilation
1   Do not draw or blow the contaminated air towards the face of the worker.
2   Place the extract as close to the source of pollution as possible.
3   Enclose as much of the source as is consistent with the work process.
4   Direct dilution ventilation so that the source of emission is entrained away from the occupied areas.
5   Discharge polluted air in such a way that it does not re-enter the building or adjacent buildings. High discharge stacks improve dispersion and minimize weather effects.
6   Make allowances for outside wind effects to prevent blowback of extracted air.
7   Make allowances for buoyancy effects of the release gases or vapours, i.e. hotter or less dense substances tend to rise when in a concentrated form, whilst colder, heavier substances tend to fall. There is also a tendency for a concentrated pollutant released into still or slow moving air to form a layer on the floor or in the roof, according to its density relative to air, the removal of which is difficult and which can lead to dangerous accumulations if the material is toxic or inflammable.
8   If it is important to contain the polluted air in the room in which it is released, this can be achieved by extracting 10–15% more air than supplied.
9   Do not discharge toxic or harmful substances into the atmosphere without rendering them harmless.

---

10   Certain gases and vapours are inflammable and their handling may come under special regulations or codes requiring flameproof equipment. Due regard must be taken of this.

Legal requirements
Specific mention of ventilation of the workplace is made in the following Acts and regulations, although many others call for ventilation.
Health and Safety at Work, etc. Act 1974
Offices, Shops and Railway Premises Act 1956, Section 7
SI 1956 No. 1764, Coal and Other Mines (Ventilation) Regulations
SI 1956 No. 1771, Coal and Other Mines (Locomotive) Regulations
SI 1972 No. 917, The Highly Flammable Liquids and Liquefied Petroleum
    Gases Regulations
SI 1988 No. 1657, Control of Substances Hazardous to Health Regulations
    1988

9.4   Choice of fan

Fan pressures
The Fan Manufacturers Association have agreed on the following definitions for fan pressure:
Fan total pressure: the rise in total pressure across the fan which is equal
    to the total pressure at the fan outlet minus the total pressure at the inlet
Fan velocity pressure: the velocity pressure based on the mean velocity at
    the fan outlet
Fan static pressure: the fan total pressure minus the fan velocity pressure.

Power and efficiency definitions
Air power (total): the theoretical power required for a volume of air to
    move against a resistance, calculated from

$$P_a = Q.p_t$$

where P_a is the air power (total), Q is the volume flow rate and p_t is the total pressure required to move that flow rate. If Q is in m s^{-1} and p_t in Pa, then the resulting power will be in watts
Fan efficiency (total): the ratio of the air power total to the input power at
    the shaft of the fan impeller
Fan power (total): the power required at the shaft of the fan impeller to
    move the air against the resistance. It is calculated from

9.4   Choice of fan

$$P_t = \frac{Qp_t}{\eta}$$

where P_t is fan power (total) and η is fan efficiency (total).

Fan characteristic curves
If a fan is run at a constant speed and its volume flow rate altered by
varying the resistance against which it has to operate, curves showing the
variation of pressure, power and efficiency can be plotted against volume
flow rate. These curves, known as fan characteristic curves, give the
performance of the fan over the whole range of resistance for which it is
designed. Manufacturers' catalogues quote these curves either as graphs
or as tables. The shape of the curve depends upon the geometric shape of
the fan.

Fan types

Propeller fan
The propeller fan is useful for general ventilation, where there is little
resistance to air flow, but is *not* suited for use in a ductwork or air
filtration, where resistances are high. It is commonly used in unit heaters,
air cooling and heat rejection devices, e.g. vehicle radiators, and is also
mounted in windows and apertures in walls or buildings, to provide general
ventilation. Propeller fans have sheet steel blades and so are low in
efficiency but, if fitted with aerofoil-shaped blades, can produce
efficiencies of up to 70%.

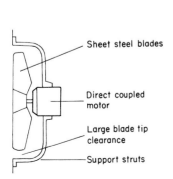

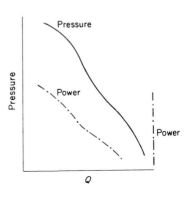

9.4   Choice of fan

Axial flow fan
In an axial flow fan, the impeller is contained within a cylindrical casing, on a shaft whose axis is along the centre-line of the casing. It is similar in principle to the propeller fan but, since the blades are of aerofoil shape and have their tips running close to the casing, they are capable of producing higher pressures, normally up to 1100 Pa, per stage. For higher pressures, a two-stage axial flow or a centrifugal fan would be required. The performance of axial flow fans depends upon the angle of the blades to the direction of rotation (see figure). Increasing the angle will increase the volume flow rate and the pressure developed but obviously more power will be required to do so. With some fans, the blade angle can be changed by unclamping the hubs and resetting them to the new angle. On the most sophisticated fans, the angle can be changed whilst in motion.

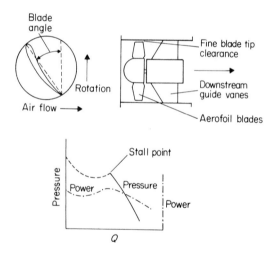

As the impeller rotates, the air is given a swirl of spiral rotation, which is undesirable from the point of view of performance. Some form of flow straightener is, therefore, required to recover some of the energy of rotation. Where the fan is to be used close to a finned device, such as a heat exchanger, the flow will be straightened by the fins but, if ducting follows the fan, some form of static vanes is required for this purpose. Many axial flow fans have static guide vanes fitted as an integral part of the construction, often acting as the support for the shaft or the motor. Some two-stage axial flow fans are arranged so that the second impeller rotates in the opposite direction to the first, thus counteracting the swirl. At the most efficient part of the characteristic axial flow, fans can produce efficiencies of up to 78% and they are very compact, often being the same diameter as the ducting to which they are connected.

The disadvantages of the axial flow fan are as follows:
• they are noisy when running at high speeds and, therefore, require silencing
• many have their motors in the airstream, which limits their use to cool, uncorrosive airstreams
• if the motors are installed outside the airstream, they lose their compactness
• they are liable to stall when the resistance against which they are expected to work is too high. This can damage the fan bearings if allowed to continue too long.

Bifurcated axial flow fan
This fan possesses most of the features of the axial flow fan but, although the impeller is directly coupled to the motor, the airstream does not pass over the motor because of the bifurcated shape of the casing. Hot and dirty air can, therefore, be handled by this fan and it is commonly used for extract ventilation.

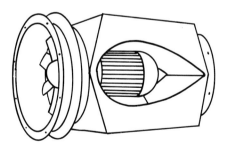

Centrifugal fans
With centrifugal fans the air enters the centre of an impeller shaped like a paddle wheel and is discharged at the periphery, into a volute-shaped casing with a single rectangular outlet. Thus, the air leaves the fan at right angles to the direction in which it entered. The blades of the impeller can

be inclined either forwards or backwards in relation to the direction of
rotation or they can be radial, each type having its own characteristics.
Irrespective of shape, centrifugal fans are capable of producing high
ventilation pressures.

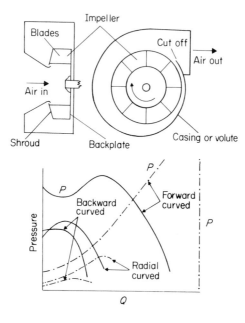

Forward-curved centrifugal fan
The impeller of a forward-curved centrifugal fan has many closely spaced
narrow blades inclined forward to the direction of rotation, giving a
characteristic curve, in which three different volume flow rates are possible
over a certain range of pressures, depending upon the resistance of the

9.4   Choice of fan

---

system against which it is operating. The power characteristic is also distinctive, in that it continues to increase with the increase in volume flow, giving an overloading feature at high flows. This means that, if the motor powering the fan is designed for some intermediate duty, it will overload if the fan is run without any resistance or if the resistance drops substantially below the design specification. This type of fan is the most compact of the centrifugal ones and is used where space is limited, such as in air-conditioning plant rooms. Efficiencies do not exceed 70% and motor and drive powers are, therefore, comparatively large.

Radial-bladed centrifugal fan (paddle-bladed)
The impeller in this fan has a few widely spaced blades disposed radially in relation to the direction of rotation. The pressure characteristic is similar to the backward-bladed fan but the power characteristic has its maximum power at the maximum volume flow rate. The blades are flat and radial and so are self-cleaning. Since they have no shroud or backplate, they are easy to replace when worn. They are, therefore, ideally suited to the handling of dirty and corrosive air. Efficiencies do not exceed 60%, so their use should be limited to places where their special features are suited.

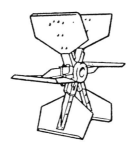

Backward-curved centrifugal fan
In this fan, the impeller has a few widely spaced blades inclined backward to the direction of rotation. They are somewhat larger than in the forward-curved centrifugal fan but they are the most efficient. When fitted with aerofoil section blades, they can develop efficiencies up to 88%. They are, thus, most popular for use in high-powered, continuously running situations, such as power stations and mines.

9.4   Choice of fan

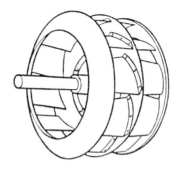

Matching of fan and system

The most efficient part of the fan pressure characteristic is usually just to the right of the peak pressure on the top of the final downward slope. A fan should, therefore, be chosen so that the calculated duty point lies on the most efficient part of its characteristic. Where it is required to know the effect of installing a particular fan on a system of known resistance, it is necessary to plot the system characteristic curve on the same ordinates as the fan curve. The point of intersection of the two curves will indicate the operating point of that fan on that system (Fig. 9.6). The volume flow rate and the pressure loss can also be read, as can the power required (from the power curve).

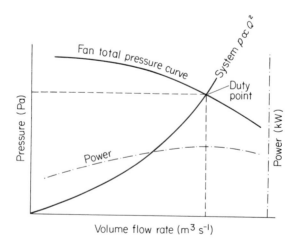

Fig. 9.6 Fan total pressure curve.

9.5   Air cleaning and discharges to the atmosphere

Air cleaning

Environmental pressures are already put upon employers to limit the amount and concentration of toxic substances discharged to the atmosphere and certain registered works are required by law to do so. These pressures will be reinforced by a more wide-ranging law during the 1990s. Therefore attention should be turned now to limiting the emissions from ventilation systems that contain substances hazardous to health.

Removal of particles from discharged air involves various alternative principles, the choice of which is dictated by the size, nature and concentration of the particles. Large dry particles in a high concentration, from, for example, a woodworking shop, are best dealt with by a centrifugal method such as a cyclone. Small dry particles, from, for example, a lead smelter, would best be removed by a fabric filter such as a bag house. Dusts that can be easily charged electrically can be removed by electrostatic means. Sticky or wet dusts will require methods involving wet scrubbing, which, although cleaning the discharged air, may lead to a secondary problem of wet sludge disposal.

Removal of gases and vapours from discharged air is generally more difficult as the pollutant is in molecular form and does not respond to the physical forces that can capture airborne particles. Chemical adsorption and absorption are the principles employed. Wet absorption involves passing the gas-laden air through extended wetted surfaces and sprays of water or chemical solutions. Adsorption normally uses activated charcoal beds through which the gas-laden air passes, the pollutant being adsorbed on to the minute pores in the charcoal. To achieve a high degree of cleaning, both techniques are bulky and costly to purchase and operate.

All air cleaning techniques absorb energy from the extracted air. This has to be taken into account when choosing the correct fan for the job. Adding air cleaning later will invariably involve upgrading the performance of the fan to cater for the extra resistance to air flow that the device provides.

Discharges to the atmosphere

Any ventilation systems that discharges air horizontally to the atmosphere will, at some time or other, encounter external wind pressures. Sometimes the direction and velocity of the outside wind will result in the air flow in the ventilation system reversing and pollutants will be scattered internally rather than be extracted. Such systems require weather covers to prevent blowback but these devices can result in polluted air remaining close to the buildings and being re-entrained into them. It is far better to discharge vertically as high as permissible with a high efflux velocity. Conical weather covers are to be discouraged as they tend to bring pollutants down to ground level to be re-entrained into adjacent buildings. Suitable alternatives are available to prevent rain and snow from entering the discharge stack.

9.6   Energy and cost implications of ventilation systems

The energy required to provide ventilation can be divided into that required to overcome air friction in the hardware and that required to heat the replacement air, where air is discharged to the atmosphere. In most cases and at most times of the year, the latter is far more costly.

The costs of providing motive power to overcome the friction of moving air through a ventilation system can be calculated from the fan power formula given previously, applying the fuel cost figures given in Table 9.5. The fuel costs given here must be regarded as only a guide because the price paid depends upon individual contracts made with suppliers and the proximity to major supply terminals. Also, at the time this book went to press, oil and gas prices were in a state of flux. For the most recent data, the source should be consulted.

Where extracted air is discarded to the atmosphere, the energy required to provide heated replacement air will depend upon the outside air temperature and the designed inside air temperature. In Britain,

Table 9.5 Comparable energy costs (adapted from National Industrial Fuel Efficiency Service Ltd.)

Energy source		Estimated fuel cost per kWh (June 1990) (pence)	Approximate heating cost per kWh using the authors' estimation of conversion efficiencies (June 1990) (pence)
Electricity	Direct	4.7	4.7
	Off-peak	2.2	2.2
Oil	35 s	2.0	2.9
	3500 s	1.5	2.2
Gas	Natural	1.3	2.0
	Propane	1.1	1.7
	Butane	1.1	1.7
Solid fuel	Coal	0.9	1.8
	Industrial coke	1.3	2.6

9.6   Energy and cost implications of ventilation systems

meteorological records are available for many sites, from which estimates of the number of hours in a year at a particular air temperature have been made. Figure 9.7 gives such values for a typical South of England site.

In order to establish the amount of energy required to heat 1 m^3s^{-1} of outside air to inside designed conditions, the chart in Fig. 9.8 can be used in conjunction with Table 9.5.

To use the chart in Fig. 9.8: the vertical lines represent outside air temperatures and the inclined represent ones inside. From where any two intersect, the power required can be read horizontally across on the left-hand column. An example (shown on the broken lines) is: outside air 0°C, inside air 18°C, power required to heat 1 m^3s^{-1} of air is 22.5 kW. Thus, for every hour that the air is drawn in, the energy required would be 22.5 kWh. From Table 9.5, the running costs can be estimated, based upon the type of fuel used.

To use the pressure loss ordinate: the theoretical power required to overcome the frictional resistance of a system can be found by reading across horizontally from where the ventilation pressure loss line intersects

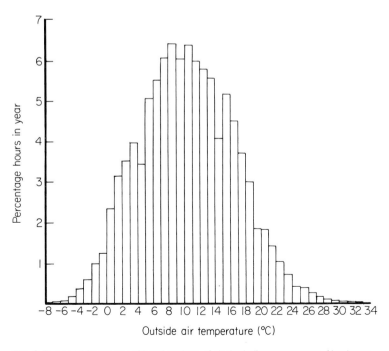

Fig. 9.7 Annual frequency of hourly values of air dry bulb temperatures, Heathrow Airport, London, 1949–76.

9.6   Energy and cost implications of ventilation systems

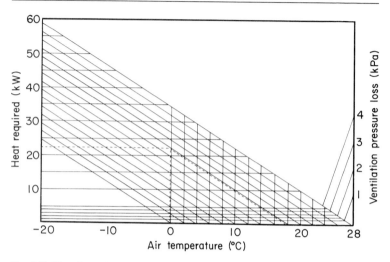

Fig. 9.8 Chart for estimating energy requirements for heating 1 m³ s⁻¹ of air from outside to inside air temperature, also showing energy equivalent of ventilation pressure loss (after Gill, *Annals of Occupational Hygiene* 1980, 23, p. 243).

with the top inclined outside temperature line. Example: a ventilation system whose pressure loss is 2 kPa for a flow rate of 1 m³s⁻¹ will give a theoretical power requirement of 2.5 kW. Thus, for every hour that it is operated, it would require a theoretical power of 2.5 kWh. The true power would have to take into account the efficiency of the fan and, if the fan was electrically driven, the cost could be obtained from Table 9.5.

10 Personal protection of the worker

10.1 Introduction, 273

10.2 Eye and face protection, 275

10.3 Skin and body protection, 278

10.4 Respiratory protection, 282

10.5 Hearing protection, 289

271

10.1 Introduction

If control at source or during transmission is not possible or if extra safeguards are required, then the worker must be protected. This can be done in one of three ways:

• by washing the worker in a stream of uncontaminated air, thus displacing any airborne pollutants
• by segregation or separation, using a shield or conditioned enclosure
• by providing personal protective clothing.

Displacement

This depends upon the creation of a diffused clean air flow over the worker and towards the work, carrying away the work products, preferably towards some form of extraction system. The provision of work station supply ventilation has corresponding economies in the volume of air required, in comparison with a system ventilating the whole building. Care must be taken to ensure that local turbulence is minimized, so that effectiveness of the control is maintained. This technique is most suited to well-defined work stations.

Thermal comfort of the worker must be considered, so that the combination of air temperature and velocity is such that cold draughts are not experienced at the work station. To this end, supply diffusers must be carefully chosen and the supply air temperature accurately controlled, to suit the air velocity directed at the worker. This is particularly important where air is discharged from above and behind the worker, as the back, the neck and the back of the head are the parts of the body most sensitive to draughts. It is important to remember that even air at a temperature above that of a normal room can feel cold if it is flowing at a sufficiently high velocity. Figure 10.1 shows the relationship between air velocity and temperature and the part of the body affected.

Figure 10.1 shows the number of people complaining of discomfort as a percentage of the total number tested. As an example, it can be seen that a draught of 0.2 m s^{-1} at a temperature of 1 °C below ambient room temperature, blowing on the occupants' necks, will give a feeling of uncomfortable coolness to 20% of the room occupants, whereas the same draught blowing on the ankle region will cause discomfort to only 5%.

Enclosing or shielding the worker

Isolating the worker from an uncongenial or toxic environment is a technique which is often adopted where the working process is too large or too expensive to control at source or in the transmission stage. Isolation

273

10 Personal protection of the worker

10.1 Introduction

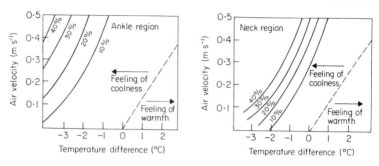

Fig. 10.1 Percentage of room occupants objecting to draughts on the ankle and neck region (Study by Houghten F.C., Gutberlet C. and Witkowski E. *ASHVE Transactions* 1938, 44, p. 289).

cubicles can be used to protect from noise, ionizing radiation, heat and cold, as well as from airborne toxins. In most cases, the enclosure will require ventilating and, possibly, air-conditioning and the amounts of air required will need to be calculated. As a general rule, each person enclosed will require 10 l of fresh air per second but this amount can be varied, depending upon the size of enclosure and whether or not smoking is permitted. For example, a small enclosure containing one person who smokes would require a fresh air rate of $25 \, l \, s^{-1}$, whereas a spacious enclosure in which no smoking takes place could be ventilated with as little as $5 \, l \, s^{-1}$ per person.

Personal protection

The organs of the human body that are vulnerable to attack from external sources are the eyes, the ears, the skin and the respiratory system. In the case of the first three of these, a barrier or attenuation device should be worn over the organ being protected. With regard to airborne pollutants, respiratory protection involves the wearing of a device that either cleans the polluted air to a safe level or provides a stream of uncontaminated air from a separate source.

At this juncture it must be pointed out that Regulation 7 of the COSHH Regulations specifically states that control should be secured by measures other than by the provision of personal protective equipment (PPE) but where other means are not preventive or do not provide adequate control, then, in addition to those measures, suitable PPE shall be provided adequately to control exposure. The Regulations require that the PPE provided must be suitable for the purpose and conform to a standard approved by the Health and Safety Executive.

Examples of situations where the use of PPE may be necessary:
- where it is not technically feasible to achieve adequate control by other measures alone then control should be achieved by other methods as far as is reasonbly practicable and then in addition PPE should be used
- where PPE is required to safeguard health until such time as adequate control is achieved by other means
- where urgent action is required such as in a plant failure and the only practical solution is to use PPE
- during routine maintenance operations.

(Source: COSHH Approved code of Practice, paragraph 33.)

In most cases, PPE appears to offer a cheap alternative solution to providing engineering control methods. On closer scrutiny, however, the management problems that are created by its introduction make this alternative less attractive. In order to make the decision routinely to use personal protective devices, it is necessary to hold discussions involving trade unions or workers' representatives and, once the decision to go ahead has been made, arrangements must be put in hand for the education and training of the users. A complete back-up system of purchasing, storage, cleaning, repair, inspection, testing and replacement has also to be established. Moreover, the reaction of the worker to being asked to wear the devices may involve financial inducements and changes of contract conditions before adoption, and management supervision may have to be strengthened.

European legislation
Late in 1989 the Council of European Communities published two important Directives regarding PPE. The first (89/656/EEC) sets out the minimum health and safety requirements for the use by workers of personal protective equipment in the workplace and requires member states to implement it by 31 December 1992. Details of this Directive can be seen in the Official Journal (OJ) of the European Communities No. L 393/19. The purpose of the second (89/686/EEC) is to harmonize the European standards of PPE through the European Committee for Standardization (CEN). This Directive can be found in OJ No. L 399/19 and is required to be implemented by 31 December 1991.

10.2 Eye and face protection
Protection must be provided to guard against
- impact of small particles projected at a low velocity
- the impact of heavy particles at a high velocity
- the splashing of a hot or corrosive liquid

• the contact of the eyes with an irritating gas or vapour
• a beam of electromagnetic radiation of various wavelengths, including laser beams.

   Each harmful agent may require a particular form of eye protection, which may be unsuitable for another agent. In some cases, the protection may have to be extended to the whole face. Whatever hazard or hazards exist, the protective device must be carefully chosen to suit.

Legal requirements
Health and Safety at Work etc. Act 1974
S1 1974—No.1681, Protection of Eyes Regulation made under Section 65 of the Factories Act 1961, requiring employers to provide eye protection and shields to persons employed in specified processes, and embracing many industrial situations.

Standards
Various standards for eye protection have been produced to assist in obtaining the correct specification to suit the harmful agent.

British Standards Institution (London)
BS 2092, Specification for Industrial Eye Protectors (1987)
BS 1542, Equipment for Eye, Face and Neck Protection Against Radiation Arising During Welding and Similar Operations (1982)
BS 679, Filters for Use During Welding (1977)
BS 2724, Filters for Protection Against Sun Glare (for General and Industrial Use) (1987)
BS 4110, Specification for Eye Protectors for Vehicle Users (1979)
BS 6967, Glossary of Terms for Personal Eye Protection (1988)
BS 7028, Selection, Use and Maintenance of Eye Protection for Industrial and Other Use (1988)

American National Standards Institution (New York)
ANSI 287.1, Standard Practice for Occupational and Educational Eye and Face Protection (1968)

General points
Eye protection takes the form of spectacles, goggles or face shields, all of which are available from a wide range of manufacturers and suppliers, in a wide range of sizes. Suitability for the hazard and comfort must be the overriding factors in choosing the particular device, as the users must have complete confidence in the protection it provides and must not be forced to remove it to relieve discomfort during the operation for which protection

is required. A preoccupation with discomfort may also distract from the task in hand and lead to errors and accidents. A good range of suitable forms of protection should, therefore, be made available for the user to choose the one to suit the shape of his/her face. This may mean having products from more than one manufacturer available.

Some of the problems involved in the use of eye protectors are given below. Several can be overcome by suitable selection but certain problems are inherent in the use of such devices.

1 They may not guard against the hazard.

2 They may not fit properly.

3 They may be uncomfortable due to uneven pressure on the face.

4 They will restrict the field of view.

5 Spectacles worn for correction of vision may interfere with the wearing of eye protectors and vice versa. Whilst safety spectacles with corrective lenses are available, their suitability is limited to minor eye hazards.

6 Optical services and follow-up may be necessary to deal with problems of refraction of light.

7 Eye protectors may interfere with the wearing of respiratory and/or hearing protection. Where more than one organ is to be protected, an integrated combined protective device may, therefore, be more suitable.

8 Due to discomfort, the wearer may be tempted to remove the protector from time to time, with a consequent loss of protection for that period.

9 Fitting, cleaning, inspection and replacement procedures are necessary.

10 Training may be required for users and for maintenance staff.

Types available

Safety spectacles are only suitable for low energy hazards but are available in a wide range of sizes to suit the face. Types: clear, clip on, prescription, tinted (anti-flash).

10.3 Skin and body protection

Goggles are suitable for a wide range of hazards but limited in fittings from any one manufacturer. Types: chemical, dust, gas, gas welding, general purpose, molten metal.

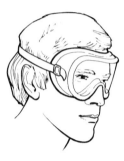

Shields are suitable to protect the eyes or the whole face; they can be attached to a helmet or a head band but may be hand-held. Types: eye, face, furnace viewing, welding.

10.3 Skin and body protection

Skin protection includes guarding hands, feet and body against
• damage from dermatitic or corrosive agents
• absorption into the body via the skin
• radiant heat
• cold
• ionizing and non-ionizing radiation
• physical damage.

The material used for the gloves, apron or garment must be suited to the purpose and must be chosen carefully.

10   Personal protection of the worker

10.3 Skin and body protection

Legal requirements

Health and Safety at Work etc. Act. 1974

SI 1950 No. 65, The Pottery (Health and Welfare) Special Regulations

SI 1948 No. 1547, The Clay Works (Welfare) Special Regulations

SI 1980 No. 1248, The Control of Lead at Work Regulations

SI 1985 No. 1333, Ionizing Radiation Regulations

SI 1988 No. 1657, The Control of Substances Hazardous to Health
  Regulations

Standards

The following British Standards apply to protective clothing.

Hand protection

BS 1651, Industrial Gloves (1986)

BS 697, Rubber Gloves for Electrical Purposes (1986)

BS 1884, Specification for Rubber Post Mortem Gloves (1952)

BS 2606, X-ray Protective Gloves for Medical Diagnostic Purposes up to
  150 kV Peak (1955)

Foot protection

BS 1870, Part 1, Specification for Safety Footwear Other than All-rubber
  and All-plastic Moulded Types (1988)

BS 1870, Part 2, Specification for Lined Rubber Safety Boots (1986)

BS 1870, Part 3, Specification for Polyvinyl Chloride Moulded Safety
  Footwear (1981)

BS 953, Methods of Test for Safety and Protective Footwear (1979)

BS 4676, Gaiters and Footwear for Protection against Burns and Impact
  Risks in Foundries (1983)

BS 4972, Ladies Protective Footwear (1973)

BS 5145, Specification for Lined Industrial Vulcanized Rubber Boots (1984)

BS 5451, Specification for Electrically Conducting and Antistatic Footwear
  (1977)

BS 5462, Footwear with Midsole Protection (1984)

BS 6159, Polyvinyl Chloride Boots (1987)

Body protection

BS 2653, Protective Clothing for Welders (1955)

BS 3791, Clothing for Protection Against Intense Heat for Short Periods
  (1970)

10.3 Skin and body protection

BS 4724, Resistance of Clothing Materials to Permeation by Liquids
   Part 1, Method of the Assessment of Breakthrough Time (1986)
   Part 2, Method for the Determination of Liquid Permeation after
   Breakthrough (1988)
BS 7182, Specification for Air-impermeable Chemical Protective Clothing
   (1989)
BS 7184, Recommendation for Selection, Use and Maintenance of
   Chemical Protective Clothing (1989)

Types available

Hand protection

Materials	Protection	Gloves
Asbestos	Abrasion	Armoured
Cotton	Chemical	Chainmail
Leather	Electrical	Disposable
Moleskin	Fire/flame/heat resistant	Electrician's
Neoprene	General purpose engineering	Gauntlets
Nitrile	Hygiene	Hand pads
Nylon	Low temperature	Mitts
Polythene	Radiation	Reversible
PVA		Surgical
PVC impregnated cotton		X-ray
PVC		
Rubber		
Terrycloth		
Terylene		

Foot and leg protection

Anti-static footwear	Knee pads
Boots and shoes	Leggings
Chemical footwear	Moulded footwear
Clogs	Non-slip footwear
Cold-storage footwear	Over boots and over shoes
Conductive footwear	Rubber ankle boots
Foundry boots	Soles and heels
Gaiters/spats	Thigh boots
Knee boots	

10.3 Skin and body protection

Body protection

Materials	Protection	Garments
Asbestos	Buoyant	Aprons
Chainmail	Chemical	Armlets and sleeves
Cotton (denim, etc.)	Exposure	Capes
Glass fibre	Fire/flame/heat-resistant	Coats and jackets
Leather	High-visibility fluorescent	Disposable gloves
Melton	Ionizing radiation	Hoods and sou'westers
Moleskin	Proofed	Overalls
Neoprene	Quilted	Suits—hot entry
Nylon, terylene	Ventilated	Trousers
Paper and disposable		
Plastic-coated		
Polyurethane		
PVC		
Wool		

General points

1   Protective clothing materials may be attacked and degraded by contact with chemicals. Protective clothing designed to withstand chemicals comes in a variety of forms and materials, each with its own characteristics with regard to resistance to chemical permeation. Permeation rates with different manufacturers' garments may vary even with the same material. Therefore it is important to know the breakthrough times before selection.

2   Whilst the material may be suitable, seams and joints in garments may allow the passage of particles, liquids and/or vapours. This can be aggravated by the bellows action of body movement within a clothing assembly.

3   Protective clothing, particularly a whole-body garment, sets up a microclimate inside which the loss of body heat may be limited, causing discomfort and leading to possible stress. Some such garments can be ventilated.

4   Some garments restrict the movement of limbs, which slows the worker and increases fatigue.

5   Provision must be made for changing, cleaning and storage of protective clothing.

6   Impervious gloves must be sufficiently long to tuck under a sleeve to prevent materials from spilling inside.

7   Low temperatures may make certain plastic materials too stiff to be usable.

10.4 Respiratory protection

The choice of equipment in this field is vast, ranging from the simple disposable dust mask to the full self-contained breathing apparatus and there is much confusion as to which device to use for a particular hazard. As the wrong choice may seriously affect the health of the wearer and could lead to asphyxiation, advice from an expert is required. Also, user training is essential, whatever device is chosen, and servicing and cleaning facilities must be provided.

Legal requirements
Health and Safety at Work Act, 1974

Factories Act, 1961, Section 30

SI 1956 No. 1768, Coal and other Mines (Fire and Rescue) Regulations (1966)

SI 1961 No. 1345, The Breathing Apparatus, etc. (Report on Examination) Order

SI 1987 No. 2115, The Control of Asbestos at Work Regulations

SI 1980 No. 1248, The Control of Lead at Work Regulations

SI 1950 No. 65, as amended by S1 1963 No. 879 and S1 1973 No. 36, The Pottery (Health and Welfare) Special Regulations

SI 1960 No. 1932, The Shipbuilding and Ship-repairing Regulations

SI 1988 No. 1657, Control of Substances Hazardous to Health Regulations

Standards
Most respiratory protection is made to conform to one of the British Standards given below.

BS 2091, Respirators for Protection against Harmful Dusts, Gases and Scheduled Agricultural Chemicals (1969)

BS 4667. Part 1 Close Circuit Breathing Apparatus (1974)

BS 4667. Part 2 Open Circuit Breathing Apparatus (1974)

BS 4667. Part 3 Fresh Air Hose and Compressed Air Line Breathing Apparatus (1974)

BS 4667. Part 4 Escape Breathing Apparatus (1982)

BS 4275, Recommendations for the Selection, Use and Maintenance of Respiratory Protective Equipment (1974)

BS 4555, High Efficiency Dust Respirators (1970)

BS 4558, Positive Pressure Powered Dust Respirators (1970)

BS 4771, Positive Pressure Powered Dust Hoods and Blouses (1971)

BS 6016, Specification for Filtering Facepiece Dust Respirators (1980)

BS 6927, Glossary of Terms for Respiratory Protective Devices (1988)

10.4   Respiratory protection

BS 6928, Classification of Respiratory Protective Devices (1988)

BS 6929, Nomenclature of Components for Respiratory Protective Devices (1988)

BS 6930, List of Equivalent Terms for Respiratory Protective Devices (1988)

BS 7004, Specification for Respiratory Protective Devices: Self-contained Open-circuit Compressed-air Breathing Apparatus (1988)

Types available

The efficiency of respiratory protection in the removal of contaminants is expressed as the nominal protection factor (npf), which is defined as the ratio of the concentration of the contaminant present in the ambient atmosphere to the calculated concentration within the facepiece, when the respiratory protection is being worn, i.e.

$$npf = \frac{\text{concentration of contaminant in the atmosphere}}{\text{concentration of contaminant in the facepiece}}$$

The nominal protection factor is used to determine the degree of protection required, knowing the concentration of pollutant in the workplace and the required concentration inhaled by the worker. Actual protection factors are often much lower than npf.

Respirators

These operate by drawing the inhaled air through a medium that will remove most of the contaminant. For dust and fibres, the medium is a filter which is replaced when dirty but, for gases and vapours, the medium is a chemical absorbent specifically designed for the gas or vapour to be removed. The medium is carried in a canister or a cartridge for ease of handling and renewal. Extreme caution must be observed to ensure that the correct medium is used for the pollutant in question and, where dust and fibres are concerned, it is important to consider the size range of the particles to be removed, in order to select the appropriate filter medium. Filters are also available for combinations of dust, gases and vapours. It is important to note that respirators do not provide any protection from an atmosphere deficient in oxygen.

Disposable respirators are manufactured from the filtering material; some are suitable for respirable size dust. The facepiece is at a negative pressure as the lung provides the motive power. npf: 5

10.4 Respiratory protection

Half-mask respirators, manufactured from rubber or plastic and designed to cover the nose and mouth; they have a replaceable filter cartridge. With the appropriate cartridge fitted, they are suitable for either dust or gas or vapour. The facepiece is at a negative pressure, as the lung provides the motive power. npf: 10

Full-facepiece respirators, manufactured from rubber or plastic, are designed to cover the mouth, nose and eyes. The filter medium is contained in a canister directly coupled or connected via a flexible tube. With the appropriate canister fitted, it is suitable for either dust or gas or

vapour. The facepiece is at a negative pressure as the lung provides the motive power. npf: 50

Powered respirators, with a half mask or full facepiece, are made of rubber or plastic maintained at a positive pressure as the air is drawn through the filter, by means of a battery-powered fan. The fan, filter and battery are normally carried on the belt, with a flexible tube to supply the cleaned air to the facepiece. npf: 500

Powered visor respirators have a fan and filters carried in a helmet, with the cleaned air blown down over the wearer's face inside a hinged visor. The visor can be fitted with side shields, which can be sized to suit the wearer's face. The battery pack is normally carried on the belt. A range of filters and absorbents are available and a welder's type is also available. npf: 1–20

There are many variations in the types of all the above devices and manufacturers' catalogues and advice should be sought before choosing.

Breathing apparatus
These provide a supply of uncontaminated air, from a source which either is drawn from fresh air or compressed air or is supplied from a high-pressure cylinder carried by the wearer.

Fresh air hose apparatus A supply of fresh air is fed to facepiece, hood or blouse, via a large-diameter flexible tube. The motive power is provided by either a manually or an electrically powered blower, giving a positive pressure in the facepiece. It is important to establish a suitable fresh air base for the blower and, if manually operated, two operators should be present. npf: 50

Compressed air-line apparatus This supplies air via a reducing valve to a facepiece, hood or blouse. If the normal factory compressed air supply is used, it is necessary to filter out contaminants such as oxides of nitrogen, carbon monoxide and oil mists from the air before introducing it to the wearer. Specially designed air compressors for breathing apparatus are preferred, as these use special lubricating oils to minimize air contamination. npf: 1000

10.4 Respiratory protection

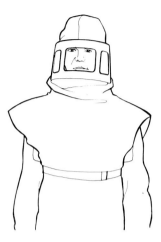

Self-contained breathing apparatus use cylinders of air or oxygen, feeding a mouthpiece or facepiece via a pressure-reducing valve. Open-circuit sets contain sufficient air or oxygen for a duration of use of between 10 and 30 min. Closed-circuit sets, which recirculate and purify exhaled breath, can last up to 3 hours. npf: 2000

General points
With negative pressure facepieces, the success of the device depends upon the adequacy fo the seal between the wearer's face and the edge of the facepiece. Sizes and shapes of human faces vary so widely that it is

important to have a range of sizes available to suit every wearer. Unfortunately, this type of respirator is unsuitable for men who wear beards, even with as little as two days' growth. British Standard 5108 suggests that the following negative pressure test will reveal leaks both around the facepiece and elsewhere in the device:

To ensure proper protection, the facepiece fit should be checked by the wearer each time he puts it on. This may be done in this way: Negative pressure test. Close the inlet of the equipment. Inhale gently so that the facepiece collapses slightly, and hold the breath for 10 seconds. If the facepiece remains in its slightly collapsed condition and no inward leakage of air is detected, the tightness of the facepiece is probably satisfactory. If the wearer detects leakage, he should re-adjust the facepiece, and repeat the test. If leakage is still noted, it can be concluded that this particular facepiece will not protect the wearer. The wearer should not continue to tighten the headband straps until they are uncomfortably tight, simply to achieve a gas-tight face fit.

Respiratory protection is generally uncomfortable to wear, particularly those forms which use the lungs to provide the motive power and have negative pressure facepieces. As the resistance of the filter has to be overcome by the wearer's lungs, the higher the resistance is, the less comfortable the apparatus becomes and the greater the temptation to remove it for some temporary respite. It has been shown that the removal of the respirator, even for a short period of time, can seriously reduce the degree of protection given—the higher the npf, the more pronounced this reduction.

In order to maintain the nominal protection factor, all devices, with the exception of disposable ones, require cleaning and inspection after use. The manufacturer should advise on the life of the canister or cartridge, taking into account the environment in which they are being used, and this must be replaced at the interval recommended. Central maintenance procedures are preferable to allowing the wearers to service their own, as nominated responsible persons can build up an expertise on care and maintenance, apply routine tests and keep records on the respirators.

Wearer training and practice is essential, even with the simplest respiratory devices, but, with breathing apparatus, training courses must be extensive and thorough. No person should be allowed to wear a set, unless s/he is seen to to fully conversant with the apparatus and knows the procedures to adopt in cases of emergency. The mines rescue services and the fire brigades have the greatest expertise and experience in the use of self-contained non-aquatic breathing apparatus in the UK.

10.5 Hearing protection

Regulation 7 of the Noise at Work Regulations 1990 (see p. 195) requires that the exposure of employees to noise should be reduced as far as is reasonably practicable by means other than the use of personal ear protectors. However, Regulation 8 requires that suitable and efficient personal ear protectors should be made available to all employees who are likely to be exposed to daily levels of noise between the 'first' (85 dB(A)) and 'second action levels' (90 dB(A)). And, if employees are likely to be exposed to levels above the 'second' or 'peak action levels' (140 dB), the employer shall provide suitable and efficient hearing protectors which when properly worn should reduce the levels to below those action levels.

Because noise is produced in a range of frequencies, the choice of hearing protection must be based upon the measured spectrum of the noise to be attenuated. Hearing protectors are either ear-muffs, which cover the ears, or ear-plugs, which are inserted into the ear canals. Within these two groups, however, there are several subdivisions. The ear-muffs can have several degrees of attenuation, whilst the ear-plugs can be of variety of materials, both disposable and reusable. Figure 10.2 shows attenuation data for four types of hearing protection and shows the importance of frequency.

It is recommended that hearing protection should be used if the workplace noise levels cannot be reduced to below 85 dB(A). The degree

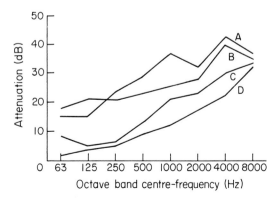

Fig. 10.2 Comparisons of the attenuation data for four hearing protectors (after Else D. (1974) in Schilling R.S.F. (ed). *Occupational Health Practice,* Butterworths, Sevenoaks). A, high-attenuation ear-muff. B, disposable expanding polyurethane foam ear-plugs. C, low-attenuation ear-muff. D, disposable ear-plugs.

of protection provided should be such that the level at the worker's ears is below 85 dB(A).

Legal requirements
Health and Safety at Work etc. Act, 1974
SI 1989, Noise at Work Regulations

Standards
BS 5108, Method of Measurement of Attenuation of Hearing Protectors at Threshold (1983)
BS 6344, Industrial Hearing Protection, Part I: Specification for Ear-muffs (1984)
    Part 2: Specification for Ear-plugs (1988)

Types available

Ear-muffs
These consist of a cup-shaped cover over each ear, held in place by a spring-loaded head-band. To ensure a good seal around the ear, the cups are edged with a cushion filled with liquid or foam. The degree of attenuation is affected by the material of the cup and its lining and the success of the device depends upon the quality of the seal around the ear.

Ear-plugs
These can be of a variety of materials.

Disposable plugs	Reusable plugs
Glass down	Permanent moulded plastic
Plastic-coated glass down	Paste-filled rubber
Wax-impregnated cotton wool	Paste-filled plastic
Polyurethane foam	

   All resuable plugs require washing after use and a sterile place for storage. Disposable plugs are available commercially in wall-mounted dispensers or in cartons containing several days' supply for one person.

Theoretical protection
To calculate the degree of protection given by hearing protectors it is necessary to measure the sound spectrum of the noise emitted at the

---

workplace, using octave band anaylsis. If the result is required in dB(A), the A-weighting values at each mid-octave frequency should be subtracted from the measured sound to provide a 'corrected level'; then each mid-octave level from that can be added together, according to Fig. 6.8. The assumed protection of the hearing protector, also expressed in mid-octave values, should then be subtracted from the corrected level and the result added as before to produce the estimated dB(A) at the wearer's ear. This can best be illustrated with an example (Table 10.1).

As with respiratory protection, hearing protection can be uncomfortable, particularly if worn for long periods, as the wearer may feel enclosed and isolated and, with ear-muffs, perspiration can build up around the seals. Whilst ear-muffs provide the greatest amount of attenuation, they are easy to remove and replace. Wearers are, thus, tempted to remove them to ease discomfort. It has been shown that removal of hearing protection, even for short periods, will reduce the overall protection substantially, the effect being increasingly more pronounced as noise levels increase. This effect is illustrated in Fig. 10.3, from which it can be seen that protection giving 20 dB(A) attenuation when worn 100% of the time will only give an effective 10 dB(A) protection when worn 90% of the time. If noise levels are much in excess of 90 dB(A) protection must, therefore, be worn continuously, to maintain levels of below 85 dB(A) at the ear, averaged over the whole shift.

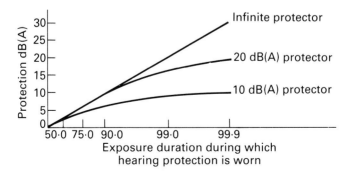

Fig. 10.3 The effects of removing hearing protectors for short periods of time. Comparisons of the protection afforded by hearing protectors which reduce the instantaneous sound level by 10 dB and 20 dB respectively (after Else D. (1974) in Schilling R.S.F. (ed). *Occupational Health Practice*, Butterworths, London).

10 Personal protection of the worker

10.5 Hearing protection

Table 10.1 Example of calculation to find the degree of attenuation provided by a particular ear-muff against a typical industrial noise

		Octave band mid-frequency Hz							
		63	125	250	500	1k	2k	4k	8k
Measured sound pressure level of typical noise	dB	92	96	102	101	98	97	94	93
A-weighting correction (see p. 191)	dB	−26	−16	−9	−3	0	+1	+1	−1
Corrected level (measured SPL minus correction)	dB	66	80	93	98	98	98	95	92
Approximate summation of levels (see Fig. 6.8)			80	99	99	101	102	97	
Estimated level of noise unprotected					104 dB(A)				
Ear-muff attenuation, mean of measurements	dB	—	13	20	33	35	38	47	41
Standard deviation	dB	—	6	6	6	6	7	8	8
Assumed protection of ear-muff (mean minus SD)	dB	—	7	14	27	29	31	39	33
Corrected level (from above)	dB	66	80	93	98	98	98	95	92
Levels at the ear (corrected level minus protection)	dB	66	73	79	71	69	67	56	59
Approximate summation of levels (see Fig. 6.8)			74	81	80	71	71	61	
Estimated level of noise at ear with protection					81 dB(A)				

10.5 Hearing protection

It can be seen from this that better overall protection may be provided by using a lower degree of attenuation from, say, glass down, which may be more comfortable and more acceptable and so not be removed during the shift, than from the higher attenuated but less comfortable ear-muffs.

Servicing and replacement facilities must be provided for ear-muffs because they will deteriorate with time, in particular at the seals, which become distorted and harden with age. A range of hearing protection should also be made available, so that wearers can choose the type that is most comfortable for them.

As with all forms of personal protective devices, adequate training must be given so that the wearers can understand the reasons for providing it. In-house training programmes should be implemented and can be aided by films and/or tape–slide presentations. Hearing protection manufacturers can assist with audio-visual aids and explanatory leaflets. Routine audiometric measurements on workers can provide an opportunity to make contact with them and encourage them to wear hearing protection.

11 Some legal background to occupational health

11.1 Sources of law in the United Kingdom, 297

11.2 European Community law, 300

11.3 International law, 303

11.4 The Health and Safety at Work Act, 304

11.5 Civil liability, 312

11.6 Medical ethics and confidentiality, 315

11.7 Industrial relations aspects of occupational health, 316

11.1 Sources of law in the United Kingdom

Acts of Paliament

Also known as legislation or statutes, these are the primary source of law in the United Kingdom. As a consequence of Parliamentary supremacy, they supersede all other types of domestic law. Bills are debated in both Houses and become law on receiving the Royal Assent. Acts come into force on the 'appointed day', but not necessarily as an entity at the same time. The Health and Safety at Work Act 1974 came into force in three stages in 1975.

Acts of Parliament remain in force until repealed. New legislation usually contains a schedule at the end containing provisions which are to be repealed. It is important to note that the Health and Safety at Work Act, unusually, did *not* repeal most of the pre-existing legislation such as the Factories Act 1961, Offices, Shops and Railway Premises Act 1963, etc. In Britain (unlike almost all other countries) there is no written constitution. As a result there is no possibility of the courts declaring legislation void, as for instance happened to the original Workers' Compensation scheme in America, which was declared unconstitutional by the US Supreme Court. British legislation is, however, now subject to European Community law, through the Treaty of Rome. It can therefore be challenged in the European Court of Justice, and must comply with EC Directives and Regulations.

An Act is said to be 'consolidating' when it brings together various pieces of legislation and amendments, as for instance the Factories Act 1961. This Act contains quite a number of specific obligations, many of them 'strict' liability independent of fault. An 'enabling Act' such as the Health and Safety at Work Act contains basically a framework of law, with precise detail left to regulations. A great deal of legislation is in this general, enabling form, nowadays.

Regulations

In terms of volume, these are by far the most important source of occupational health law. Regulations can only be made under power granted by Act of Parliament, usually to a Government Minister—in the case of health and safety regulations, usually the Secretary of State for Employment. If regulations are made outside this power, courts can declare them *ultra vires,* and thereby null and void. Regulations are not debated by Parliament, but are 'laid on the table', giving Members of Parliament a right to request a debate, which seldom happens. Regulations are examples of delegated or subordinate legislation, and are published in

11 Some legal background to occupational health

11.1 Sources of law in the United Kingdom

Statutory Instruments (SI). In the context of health and safety these are usually drafted by HSE or HSC or the Department of Employment after consultation with interested parties. Although regulations are an inferior source of law to legislation, it is occasionally possible for regulations to modify the original Act (Section 15 Health and Safety at Work Act.) Thus regulations may reduce the standard of care required by statute, or impose a different duty.

Many important new regulations have come into force within the past four years, including:

1   Arguably the most important regulations since the Health and Safety at Work Act, the Control of Substances Hazardous to Health Regulations 1988 (COSHH) (SI 1988/1657) came into force on 1 October 1989, and require employers to protect employees from 'substances hazardous to health' (other than lead, radiation, and asbestos—which are already regulated), by carrying out formal assessments, monitoring exposure, providing instruction to employees and health surveillance, including regular medical checks and keeping health records.

2   Noise at Work Regulations 1989 (SI 1989/1790).

3   Electricity at Work Regulations 1989 (SI 1989/635).

(Manual Handling Regulations are pending.)

Regulations are binding in exactly the same way as a statute, breach of which constitutes a criminal offence and, unless they state otherwise, creates civil liability.

Approved codes of practice

These are given a new status by the HSW Act (sections 16 and 17). Basically, codes of practice are not meant to have the force of law, but rather to give guidance as to correct standards. However, in *criminal* proceedings if any provision of a code appears to the court to be relevant to any matter which it is necessary for the prosecution to prove in order to establish an offence, 'that matter shall be taken to be proved unless the court is satisfied that the requirement or prohibition was in respect of that matter complied with otherwise than by way of observance of that provision of the code' (section 17 [2]).

The HSW Act is silent on the status of codes of practice in civil cases though breach of a code of practice is often held to be good evidence of negligence. On the other hand, observing a code does not of itself rule out a finding of negligence on the part of an employer.

Issuing approved codes of practice is one of the functions of the Health and Safety Commission.

11 Some legal background to occupational health

11.1 Sources of law in the United Kingdom

Guidance notes

These are found in an increasing number of documents issued by the HSC and HSE, e.g. Safety Representatives and Safety Committee Regulations. They are intended to be helpful advice but have no status in law. However, being backed by considerable technical expertise they must be regarded as of importance for they contain useful practical advice.

Case law or judicial precedent

This is basically the lawyers' speciality. Reports of decided cases are analysed to find the *ratio decidedi,* or rule of law to be followed in future cases. Precedent depends on the hierarchy of courts, i.e. courts are generally only bound by decisions of courts higher than themselves, by a principle known as *stare decisis.* The House of Lords is therefore not bound at all, and since 1966 not even by its own previous decisions. Case law is vitally important for an understanding of occupational health law. It is to the courts that we look for many of the definitions of words used in the statutes and regulations, e.g. 'factory', 'article', 'dangerous', 'reasonably practicable' have all been interpreted by judges and given a legal meaning. Sometimes these meanings are helpful to the cause of industrial safety and health, sometimes not. Deciding cases calls for striking a balance between conflicting interests, which accounts for some of its uncertainty. Only a fraction of decided cases are reported. In the UK judges tend to adopt a literal approach to the interpretation of statistics. This has put them at odds with their European counterparts, where, in interpreting the provisions of EC law, the concept of 'proportionality' applies.

Scotland and England are different countries for the purposes of judge-made law, having distinct legal systems. English decisions are not binding in Scotland and vice versa, though in civil cases Scottish appeals do go to the House of Lords, where two of the nine Law Lords are, by convention, Scottish. The principles of negligence in the two countries are more or less identical (*Donogue* v. *Stevenson*, the leading case in negligence, originated from Paisley, Scotland), but otherwise the two systems diverge quite markedly, Scots law being closer to the European civil law tradition. Most countries of the British Commonwealth and Canada (but not Quebec), Australia, India, anglophonic Africa, etc. have had their legal systems deeply influenced by English law and are still basically common law jurisdictions, albeit with major statutory differences. The law in the United States (except for Louisiana) is also basically common law in all its essentials though with two hundred years of separately developed case law, and, of course, the influence of a written constitution. Judges in the

UK occasionally look at the decisions from other common law jurisdictions, almost never from European Civil Law systems. This accords with legal tradition, albeit short-sighted.

Industrial tribunals

These have existed since 1964 and now deal with a wide range of employment law matters, particularly claims for unfair dismissal and redundancy payments. Originally intended to be a quicker and less costly form of justice than courts, they have grown in complexity in recent years. The tribunal consists of a legally qualified part-time or full-time chairman and two others, non-lawyers, drawn from either side of industry. Industrial tribunal decisions are now frequently reported, but care should be taken with these as they do not strictly speaking create a precedent, even for other tribunals. If cited they are of only persuasive authority. In employment matters there is an appeal to the Employment Appeal Tribunal, which is in fact a Court of Law and is presided over by a judge with two lay members. Decisions of the EAT are binding on industrial tribunals. Appeals against enforcement notices go to the industrial tribunal.

Orders in Council

Historically these were promulgated by the King or Queen in Privy Council, exercising the Royal Perogative. This was a convenient way of avoiding Parliamentary scrutiny, but since the Statutory Instruments Act 1946 this is no longer so, and Orders are now subject to annulment on a resolution of either House or Parliament. In modern times, power to make Orders is conferred on the appropriate Minister. For example, under power granted by section 84 of the HSW Act, the Act was extended to offshore installations and pipelines 1989 (SI 1989/840) in the Health and Safety at Work Act (Application outside Great Britain) Order 1989. Orders are, since 1972, used to apply British legislation to Northern Ireland.

11.2 European Community law

There is an increasing amount of law of the greatest importance emanating from the institutions of the EC. These mainly take the form of Directives, e.g. Dangerous Substances, Noise at Work, Product Liability. At a summit in Strasbourg in December 1989, the 'Social Charter', or Community Charter of Basic Social Rights for Workers, was signed by all member states, except the UK. The Charter will act as a focal point for a series of social programmes, based on 13 headings. These create rights and general principles for the citizens of the Community. The EC Commission's programme envisages over 40 proposals to be presented before 1992. Over half of these are to be in the form of legally binding measures, the rest comprising opinions, recommendations and memoranda. This will further add to the volume of EC law in years to come. The influence of this type of law is shown by the fact that the proportion of internationally originating health and safety legislative projects in Britain increased from 22% in 1981–82, to 67% in 1989–90.

The sources of EC law

1 Basic constitutional treaties, the most important of which is the Treaty of Rome. The Treaty of Rome's provisions are known as Articles, and are enforceable in the UK, through the European Communities Act 1972.

2 Regulations—under Article 189, these become part of member state domestic law, without the need of legislation to implement them.

3 Decisions—these can emanate from the Commission or Council of Ministers. They can be addressed to individuals, companies or other bodies, in which case they affect the recipient only.

4 Directives—these are an important course of new initiatives in EC law. They normally require national legislation to be effective. If a state ignores directives, as is not altogether uncommon, then it can be subject to infringement proceedings.

5 Judgments of the European Court of Justice. Until recently this was a court of last resort. There is now an EC Court of first instance, i.e. one that can hear cases for the first time.

6 Recommendations. These are non-binding.

The importance of EC directives

These are by far the most important source of new EC law on health and safety. The UK must implement these in its domestic legislation, if its own law does not already comply. New UK regulations on lead, asbestos, and major industrial hazards were all based on EC directives, and the Consumer Protection Act, 1987, dealing with product safety came about as a result of the EC Directive on Product Safety.

The Single European Act

An important step towards an integrated Europe was realized by the Single European Act 1988.

New Articles 8A, 8B and 8C in the Treaty of Rome specify that the EC must adopt measures to dismantle, progressively, social barriers, such as varying standards of health and safety. Thus, new Article 118A provides

that member states must pay particular attention to encouraging
improvements, especially in the working environment, with regard to health
and safety of workers. Prior to this, there had been no specific provision in
the Treaty for health and safety, and most directives in that area had been
adopted under Article 100, concerned with the 'approximation of laws
affecting the . . . functioning of the common market'. In order to facilitate
achievement of these new objectives, the Act provides for Council decisions
in these areas to be reached on the basis of a 'qualified majority vote' (QMV).

Consequently, six new draft directives were proposed in 1988, starting
with the health and safety framework directive. Under the umbrella of the
general framework directive, five 'daughter' directives were published, in
the spring of 1988, dealing with
1   Minimum workplace standards.
2   Machinery safety.
3   Visual display units.
4   Handling heavy loads.
5   Personal protective equipment.

Framework directive, and its effect on UK law
The 'framework' directive (89/391/EEC) on the introduction of measures
to encourage improvements in the safety and health of workers at work
was adopted by the EC Labour and Social Affairs Council on 12 June
1989. Its overall purpose is to establish a common set of minimum
standards throughout the Community. It does not stop member states
from imposing more onerous requirements.

The directive, by virtue of Article 2, applies to 'all sectors of activity,
both public and private' but member states are permitted to exclude
certain public services such as the police and the armed forces. It is not,
however, concerned with the protection of the public from work activities,
nor does it apply to non-employees. The scope of the directive is therefore
somewhat narrower than that of the HSW Act.

The general duty of care
Article 5 requires employers to 'ensure the safety and health of workers in
every aspect related to work'. This absolute duty is qualified in so far as
the directive allows member states to exclude or limit the responsibilities
of employers 'where occurrences are due to unusual and unforeseeable
circumstances, beyond the employers' control, or to exceptional events,
the consequences of which could not have been avoided despite the
exercise of all due diligence'. This is usually called 'force majeure'.

The effect of the qualification is clearly to bring the duty in Article 5 closer to the test to reasonable practicability laid down in the HSW Act. However, it still does not provide for the type of cost–benefit analysis which the UK courts have traditionally used when determining what was reasonably practicable in a particular situation. Nevertheless, it seems unlikely that the HSC will propose changing the duty of care laid down in section 2 of the 1974 Act.

Other changes likely to be made in order to implement the framework directive include

• regulations on the principles that employers should adopt towards risk prevention and the types of protective and preventive services they need

• the current legal framework relating to fire precautions may be amended

• the present provisions on first aid and the reporting of accidents may have to be changed

• the law on unfair dismissal may require amendment in order to provide protection for all workers who stop work in situations of 'serious, imminent and unavoidable danger'

• new duties have been imposed on employers with regard to the provision of information and training to employees, and the employees of other employers working at their workplaces

• regulations concerning co-operation and co-ordination between employers where two or more undertakings share a workplace may be introduced

• the general health and safety duties of employees are likely to be spelt out in greater detail

• the current regulatory framework relating to safety representatives and safety committees may need to be revised to ensure that all employers are required to consult over health and safety issues, though the present UK law may be sufficient.

These new regulations will take some time before they are all in force. The UK has until 1 January 1993 to implement the directive. In addition to the framework and daughter directives, there have been at least 13 other directives in progress on health and safety and related matters, such as environmental health, including the management of construction sites, safety and health signs, pregnant women at work, and the creation of a new European Safety and Health Agency. There is also a recommendation for the adoption of a new European schedule of industrial diseases.

11.3 International law

An important source of international law for occupational health is that which emanates from the International Labour Organization (ILO). The ILO

---

was founded in 1919 and consists of representatives of national government, employers and workers' organizations. Its aim is to improve international labour standards including occupational health and safety. The ILO issues Conventions which, if ratified by a state, may be treated as adopting its provisions, and also Recommendations, which do not require ratification but merely serve as a guide to national action on a particular topic. ILO Conventions and Recommendations are submitted to governments for approval, but are not automatically legally binding.

⌐The ILO also produces research papers, suggests international classification standards and issues codes of practice giving guidance on practical measures which may be taken to safeguard workers health against occupational hazards.

11.4 The Health and Safety at Work Act

Since early in the nineteenth century there has been legislation laying down criminal penalties for breaches of what came to be known as Factories Acts. When the Robens Committee looked at the situation in 1972 there were many such statutes, each dealing with different types of workplace and enforced by many different inspectorates. The Health and Safety at Work Act 1974 was an attempt to reform the legal structure in this area, and, although much of the pre-exisitng law was repealed in 1974, the Factories Act and several other notable pieces of legislation remained, with the intention of gradually replacing them in due course. This process is still continuing.

The Health and Safety at Work Act was intended to provide the following:

One system of law
One comprehensive and integrated system of law dealing with the health and safety of virtually all people at work, and also those members of the public where they may be affected by activities of people at work.

General duties
The use of broad 'general duties' as the basis for offences in the enforcement of the statute (ss. 2–9); cf. section 2: 'it shall be the duty of every employer to ensure, so far as is reasonably practicable, the health, safety and welfare at work of all his employees'. Those in control of premises (section 4), manufacturers (section 6) and employees (sections 7 and 8) may all be prosecuted under these sections.

The phrase 'reasonably practicable' as used throughout the Act was interpreted by the Courts as long ago as 1949 to mean:

'Reasonably practicable' is a narrower term than 'physically possible' and seems to imply that a computation must be made by the owner (employer) in which the quantum of risk is placed in one scale and the sacrifice involved in the measures necessary for averting the risk (whether in money, time, or trouble) is placed in the other; and if it be shown that there is a gross disproportion between them—the risk being insignificant in relation to the sacrifice of the defendants then this discharges the onus on them. Moreover the computation falls to be made by the owner (employer) at a point in time anterior to the accident
(Lord Justice Asquith in *Edwards* v. *National Coal Board* (1949) 1 K.B. 712).

These words would still apply today. 'Reasonably practicable' therefore basically involves a cost/benefit analysis of foreseeable risks before any accident or illness develops. Note, it is for the *defendant* to prove it was *not* reasonably practicable to comply with the Act (section 40). This he has to prove only on the lower civil standard of 'balance of probabilities' rather than that of 'reasonable doubt'.

Prosecutions are normally taken in the magistrates' court where the maximum fine has been increased under the Criminal Justice Act 1991 to £5000 for each offence, although the average fine is usually much less than this. Prosecutions may on occasion go to the Crown Court, where fines are unlimited (see *R* v. *Swan Hunter Shipbuilders* [1982] 1 All E.R. 264). There is a recent example of a fine of £500 000 being imposed, and two of £250 000. These are, however, exceptional.

There is also provision for prosecution of individuals where an offence is proved to have been committed with the consent or connivance of, or to have been attributable to any neglect on the part of any director, manager, secretary or other similar officer of the body corporate or a person who was purporting to act in any such capacity' (section 37). Such prosecutions are rare but may also be on indictment in the Crown Court. In the event of someone being killed lawyers have recently been debating the possibility and extent of the crime of corporate manslaughter.

The issuing of an enforcement notice does not prevent a prosecution from being initiated for the same offence, and failure to comply with an enforcement notice is itself a separate offence. Prosecutions are still often taken using the Factories Act, etc.

Creation of the Health and Safety Commission
The Health and Safety Commission is an independent Government agency created by Act of Parliament. Of its nine members, three must be drawn

from each side of industry and the rest from local authorities or other professional bodies concerned with the general purposes of the Act. The Commission carries out the following functions:
• the making of arrangements for the carrying out of research, the publication of the results of research and the provision of training and information in connection with health and safety
• the making of arrangements for ensuring that Government departments, employers, employees, organizations representing employers and employees respectively, and other persons concerned with any matters relevant to the general functions of the Commission, are provided with an information and advisory service and are kept informed of, and adequately advised on, such matters as may be significant to them. One of the Commission's primary functions is to advise the Secretary of State on the content of statutory regulations
• the submission, from time to time, of such proposals as it considers appropriate for the making of regulations under any of the statutory provisions. In addition, the Commission is responsible for approving and issuing codes of practice.

The enforcement of health and safety legislation in the UK
Numerous independent inspectorates such as factories, mines and quarries, agriculture, etc. are now within the Health and Safety Executive, although local authorities still carry out functions in environmental and occupational health law.

The Health and Safety Executive is the central body entrusted with enforcement of health and safety legislation. Enforcement powers rest with the 'enforcing authority', the general rule being that the HSE is responsible for industrial premises and the local authorities for commercial premises. Neither inspectorate can enforce provisions in respect of its own premises. Each enforcing authority is empowered to appoint suitably qualified persons as inspectors for the purpose of carrying into effect the 'relevant statutory provisions' within the authority's responsibility.

The Health and Safety (Enforcement Authority) Regulations 1989 (s. 1. 1989/1903), which came into effect on 1 April 1990, allocate responsibility between the HSE and local authorities. Other 'relevant statutory provisions' may also provide that some other body is responsible for the enforcement of a particular requirement (HSW Act s. 18(1), 7(a).

11.4   The Health and Safety at Work Act

Jurisdiction of enforcing authorities

Activities for which the HSE is the enforcing authority
The HSE is specifically the enforcing authority in respect of the following activities
1   Any activity in a mine or quarry.
2   Fairground activity.
3   Any activity in premises occupied by a radio, television or film undertaking, where broadcasting, recording, filming, or video-recording is carried on.
4   (a) Construction work (with modifications).
     (b) Installation, maintenance or repair of gas systems or work in connection with a gas fitting.
     (c) Installation, maintenance or repair of electricity systems.
     (d) Work with ionizing radiations.
5   Use of ionizing radiations for medical exposure.
6   Any activity in radiography premises where work with ionizing radiations is carried on.
7   Agricultural activities, including agricultural shows.
8   Any activity on board a sea-going ship.
9   Ski slope, ski lift, ski tow or cable car activities.
10   Fish, maggot and game breeding (but not in a zoo).
(Enforcing Authority Regulations 1989, Reg. 4(5)(b) and 2 Sch.)
     The HSE is the enforcing authority against the following, and for any premises they occupy, including parts of the premises occupied by others providing services for them:
11   Local authorities.
12   Parish councils and community councils in England and Wales.
13   Police authorities.
14   Fire authorities.
15   International HQs and defence organizations and visiting forces.
16   United Kingdom Atomic Energy Authority (UKAEA).
17   The Crown (except where premises are occupied by the HSE itself).
     The HSE is also the enforcing authority for:
18   Indoor sports activity (with conditions).
19   Enforcement of HSWA s. 6 (duties of manufacturers/suppliers of industrial products).
(Enforcing Authority Regulations 1989, Reg. 4 and 1 Sch.)

Activities for which local authorities are the enforcing authorities
Where the main activity carried on in premises is one of the following, the local authority is the enforcing authority (i.e. a district council or London borough council, or, in Scotland, the islands or district council). (Enforcing Authority Regulations 1989, Reg. 2(1).)

1   Sale or storage of goods for retail/wholesale distribution (including sale and fitting of motor car tyres, exhausts, windscreens or sunroofs), except
    (a) where it is part of a transport undertaking;
    (b) at container depots where the main activity is the storage of goods in course of transit to or from dock premises, an airport or railway;
    (c) where the main activity is the sale or storage for wholesale distribution of any dangerous substances;
    (d) where the main activity is the sale or storage of water or sewage or their by-products or natural or town gas.

2   Display or demonstration of goods at an exhibition, being offered or advertised for sale.

3   Office activities.

4   Catering services.

5   Provision of permanent or temporary residential accommodation, including sites for caravans or campers.

6   Consumer services provided in a shop, except:
    (a) dry cleaning;
    (b) radio/television repairs.

7   Cleaning (wet or dry) in coin-operated units in launderettes etc.

8   Baths, saunas, solariums, massage parlours, premises for hair transplant, skin piercing, manicuring or other cosmetic services and therapeutic treatments, except where supervised by a doctor, dentist, physiotherapist, osteopath or chiropractor.

9   Practice or presentation of arts, sports, games, entertainment or other cultural/recreational activities, unless carried on:
    (a) in a museum;
    (b) in an art gallery;
    (c) in a theatre;
    (d) where the main activity is the exhibition of a cave to the public.

10   Hiring out of pleasure craft for use on inland waters.

11   Care, treatment, accommodation or exhibition of animals, birds or other creatures, except where the main activity is:
    (a) horse breeding/horse training at stables;

    (b) agricultural activity;

    (c) veterinary surgery.

12   Undertaking, but not embalming or coffin making.

13   Church worship/religious meetings.

(Enforcing Authority Regulations 1989, Reg. 3(1) and 1 Sch.)

    A new provision states that, where premises are occupied by more than one occupier, each part separately occupied is 'separate premises', for the purposes of enforcement (Reg 3(2)). But this does not apply in the case of:

    (a) airport land;

    (b) a tunnel system;

    (c) offshore installation;

    (d) building/construction sites;

    (e) university, polytechnic, college, school, etc. campuses;

    (f) hospitals;

where HSE is the enforcing authority for the whole of the premises. (Reg. 3(1) and 1 Sch.)

    Health and safety inspectors have the following powers:

• An inspector is entitled to enter any premises at any reasonable time. If he is of the opinion that the situation in the premises may constitute a danger he may enter at any time. He has a right of access to all places subject to the 1974 Act.

• He is entitled to take with him a constable if he has reasonable cause to apprehend serious obstruction in the breach of his duty.

• He may bring with him any equipment or materials required for any purposes for which power of entry was exercised and may make such examination or investigation as may be necessary.

• He may direct that any premises or part of premises he has entered be left undisturbed for as long as is reasonably necessary for the purposes of any examination or investigation.

• He may take measurements, photographs and recordings and a sample of any article or substances found on the premises.

• He may, if he identifies any article or substance as being likely to cause danger to health or safety, have it subjected to any process or test or cause it to be dismantled.

• Any person whom an inspector has reasonable cause to believe to be able to give any information relevant to an examination or investigation may be obliged by an inspector to answer such questions as the inspector thinks fit, and to sign a declaration of the truth of his answers.

• He may require the production of any books or documents which are

---

required to be kept under any of the relevant statutory provisions. It is stressed that there are important limitations on his power of disclosure of such information.
• He is empowered to require any person to afford him such facilities and assistance with respect to any matters or things within that person's control as may be necessary to enable him effectively to exercise his powers.

Enforcement notices
Since 1975 inspectors may now issue notices. These may either be improvement notices or prohibition notices. The latter may be either 'immediate' or 'deferred'.

Improvement notice (section 21) The inspector must be of the opinion that a person (i) is contravening one or more of the relevant statutory provisions or (ii) has contravened one or more of those provisions in circumstances that make it likely that the contravention will continue or be repeated.

Prohibition notices (section 22) The inspector must be of the opinion that activities involve or will involve a risk of serious personal injury.

Appeals against enforcement notices (section 24) A person on whom a notice is served may within 21 days appeal to an industrial tribunal. On such an appeal the tribunal may either cancel or affirm the notice and, if it affirms it, may do so either in its original form or with such modifications as the tribunal may in the circumstances think fit. In the event of appeal against an improvement notice, the notice is suspended, but not in the case of an appeal against a prohibition notice.
    After the industrial tribunal has heard the appeal it may be asked within 7 days to review its own decision on the following grounds:
• error on the part of the tribunal staff
• a party did not receive notice of the proceedings
• absence of a party entitled to be heard
• new evidence has been obtained which could not have been discovered before, or the interests of justice require a review.
    There still remains the further possibility of seeking judicial review in the Divisional Court and further appeal to the Court of Appeal, if further challenge is desired.

11.4 The Health and Safety at Work Act

Employment Medical Advisory Service (EMAS)

The Employment Medical Advisory Service (EMAS) was established by the Employment Medical Advisory Service Act 1972. It provides for doctors engaged in industrial medicine to carry out selective examination of young persons, and the statutory examination of persons engaged in hazardous jobs or on hazardous tasks. The nucleus of the service is the Medical Services Division of the Health and Safety Executive (HSE). It consists of a national network of some 140 doctors and nurses accountable to nine Senior Medical Employment Advisers, and is headed by the Director of Medical Services of the HSE, who takes advice from a team of specialists.

Specific duties include
• medical advice to young persons seeking employment
• medical examinations of young persons when identified as necessary by the School Medical Service
• statutory medical examinations, e.g. lead work, chemical work, fuel rigs
• advice to the factory inspectorate
• the investigation of gassing accidents
• advice to trade unions, employers, physicians
• national and local surveys, e.g. pottery industry survey, asbestos survey
• smaller pilot surveys to identify new hazards or to assess the adequacy or otherwise of current threshold limit values.

Statutory medical examinations

Over 20 000 statutory examinations are carried out by EMAS every year. A further 90 000 examinations a year are carried out by doctors employed by companies concerned who have been appointed for the purpose by EMAS. They are termed 'appointed doctors'. The fee for these examinations is a matter for agreement between the doctors and the employers concerned.

It is prescribed by statutory requirements that
• the employer is officially informed of the fitness of a worker to undertake employment
• the employee has a duty to undergo examinations
• the employer cannot lawfully continue to employ any worker who is found to be unfit
• the employee must be removed from that particular job for the prescribed period and transferred to alternative work if it is available
• the outcome of the medical examination must be recorded in a health register kept by the employer.

Worker involvement in occupational health
This is achieved through trade union appointed representatives and the use of safety committees (see Industrial relations, Section 11.7).

11.5 Civil liability

Negligence
The tort of negligence is governed by general principles laid down by the courts, which can and do change from time to time. These principles are applicable to road accidents, medical negligence or employers' liability, or any other situation where a legal duty of care arises. This is not synonymous with a moral duty of care, but a subject of legal precedent.

Persons injured or made ill or handicapped by their conditions of work may sue the person responsible (usually their employer). There are two courses of action open to such persons:
- action based on the tort of negligence and/or
- action based on breach of statutory duty.

In negligence, the plaintiff has to prove
- the existence of a duty of care (this is usually clear in employer's liability cases)
- breach of the duty of care (did the employer act according to the standard of a 'reasonable person'?)
- show that the breach of duty caused the plaintiff's injuries and that the injury itself or type of injury was not too 'remote' from the employer's act (was the injury 'reasonably foreseeable?').

In employer's liability cases, the duty of care is usually broken down into duties to provide
- competent staff and personnel
- adequate and safe plant and equipment (the employer can be liable even if the manufacturers are at fault—see Employers Liability (Defective Equipment) Act 1969).
- safe place of work
- safe system or work.

(See *Wilsons and Clyde Coal Co.* v *English* [1935] AC. 57.)

These duties are personal to the employer and cannot be delegated. Although normally only owed to employees they may also extend to others whom it is 'reasonable to foresee' might be at risk, e.g. sub-contractors. (See *McArdle* v. *Andmac Roofing Co.* [1967] 1 All E.R. 583.)

Furthermore, employers may be liable for the negligence of their employees whose acts have injured others, including fellow workers, 'in the course of their employment'. This is called vicarious liability. Employers are not vicariously liable for the acts of independent contractors.

Recently many victims of deafness or noise-induced hearing loss have recovered damages from their employers by means of actions in the tort of negligence.

Civil actions have also been brought for breach of contract, where this is legally desirable, e.g. injuries occurring abroad.

Breach of statutory duty

Whether or not the employer is prosecuted, an employee who has suffered injury may sue his employer for breach of statutory duty. The plaintiff has to prove

• that he or she is one of the protected classes of person under the statute.
• breach of relevant health and safety statute or regulation
• that the breach caused his or her injury
• that the injury was not too remote.

Often this is easier to prove than negligence as the statute may not require fault on the employer's part, though most in fact do e.g. 'reasonably practicable', 'safe',. 'dangerous'—all require foreseeability of the risk. If the employer has already been convicted for an offence involving breach of the statute in question, the conviction may be used in evidence in subsequent civil proceedings (Civil Evidence Act 1968).

Note: The 'general duties' sections of the Health and Safety at Work Act do *not* allow a civil action, only enforcement by the inspectorate. However, a civil action may be taken for breach of health and safety regulations made under the Act, unless the regulations say otherwise (section 47 [2]). Almost all the sections of the Factories Act allow for civil action, although those of a purely welfare character do not. It is therefore important to establish at the outset the proper construction of any statute or regulation before embarking on civil litigation.

Civil actions may also be brought against manufacturers of items which cause injury to persons. This may either be in negligence or under the provisions of the Consumer Protection Act 1987 (based on the EC directive on product liability), which provides for strict liability (i.e. without proof of negligence or fault). The Consumer Protection Act 1987 also makes changes to section S.6 of the Health and Safety at Work Act as far as the criminal liability of manufacturers and suppliers, etc. is concerned.

11.5 Civil liability

Statutes of limitation

Occupationally linked disease or disability frequently takes a long time to manifest itself and sometimes plaintiffs do not proceed as quickly as they might in pursuing their claims. This sometimes results in these claims being statute-barred, which means that if their claim was not initiated within three years of the date of cause of action it cannot continue. The law has been changed in the Limitation Act 1980. Section 11 of the 1980 Act provides that in personal injury cases the limitation period is three years from either the date on which the cause of action accrued, or the date of knowledge (if later) of the person injured. Personal injury includes any disease or any impairment of a person's physical or mental condition. 'Date of knowledge' is defined in section 14 of the Act as the date on which the person concerned first knew:

1 That the injury in question was significant, and that the injury was attributable in whole or in part to the act or omission which is alleged to constitute negligence.

2 The identity of the defendant, and, where it is alleged that the act or omission is that of some person other than the defendant, the identity of that person and the additional facts supporting the bringing of an action against the defendant.

Also relevant is section 33. The courts are now empowered to override the time limits if it appears to the court that it would be equitable to do so having regard to the degree to which

• the provisions of sections 11 and 12 of the Act prejudice the plaintiff
• any decision of the court under the subsection would prejudice the defendant.

In deciding whether to exercise this discretion the court has to have regard to all the circumstances of the case and in particular to

• the length of and reasons for the delay on the part of the plaintiff
• the extent of loss of cogency of the evidence because of the delay
• the conduct of the defendant after the cause of action arose
• the duration of any disability of the plaintiff after the cause of action arose
• the extent to which the plaintiff promptly and reasonably acted once he knew he might have an action for damages
• the steps, if any, taken by the plaintiff to obtain medical, legal or other expert advice and the nature of any such advice he may have received.

It must be stressed these rules apply only to civil proceedings and not criminal prosecutions, where different principles operate.

11.6 Medical ethics and confidentiality

A host of important areas of occupational health practice raise legal issues.
The answers to these problems are too complex for a work of this sort,
but they are of such importance that the practitioner should be aware of
them and have regard to their consequences.

On medical ethics, the guidance of the BMA and the Faculty of
Occupational Medicine can have legal implications though not themselves
legally binding:

• Defining the role of the occupational health practitioner. There are
important considerations to be found in contracts of employment.
• What are the legal effects of disclaimers or exemptions from liability?
• How is causation defined in law, for example in relation to negligence?
Does this differ from the medical or scientific concept?
• Confidentiality of medical records—implications of legal rules of
discovery and disclosure.
• Implications of Data Protection Act 1984.
• Access to Medical Records Act 1988. Under this fairly new piece of
legislation individuals have a statutory right of access to medical reports
relating to themselves, prepared for employment or insurance purposes.
They are able to comment on and, where appropriate, amend or qualify the
contents of reports before they are sent, ensuring that inaccurate or
misleading information is not sent to third parties. Furthermore, the Act
entitles individuals to withhold consent for a report to be supplied or,
indeed, for an application for a report to be made in the first place. This
new legislation has significant implications for employers seeking reports
both on current and prospective employees, for example in cases of ill-
health dismissals or pre-employment health investigations.

Unlike the Data Protection Act 1984 the Access to Medical Records
Act applies not only to reports stored on computer but also to those
stored manually. It does not apply to medical records *per se;* however, a
doctor who, instead of preparing a medical report in the normal way,
decides to send either partial or complete copies of a patient's medical
records would be obliged to treat those records as if they were a report
for the purposes of the Act.

The Act only applies to reports produced by medical practitioners who
are or have been responsible for the clinical care of the individual. It does
not cover reports produced by medical advisers who do not fulfil this
criterion, and is therefore unlikely to apply to reports produced by
company doctors or independet medical advisers following a one-off
medical examination for employment or insurance purposes.

11.7 Industrial relations aspects of occupational health

Safety representatives and safety committees

These were another innovation of the Health and Safety at Work Act. The detailed regulations and code of practice came into force in 1977.

In order to secure greater co-operation among the workforce, the Act requires that safety representatives be appointed by recognized trade unions, to represent the employees in consultation with the employers on health and safety matters. If requested to do so by safety representatives the employer must set up a safety committee. It is also the employer's duty to consult with safety representatives with a view to the making and maintenance of arrangements which will enable him and his employees to ensure health and safety. The issuing (and revising) of a safety policy is but one aspect of this general duty. It is an offence not to have a safety policy.

The regulations set out the function of safety representatives. These are

* to represent the employees in consultation with the employer
* to make representations to the employer on any general or specific matter affecting the health and safety of employees
* to carry out inspections of the workplace
* to receive information from inspectors
* to inspect documents which the employer is required to keep by virtue of the relevant statutory provisions.

The code of practice encourages employers to provide information to all employees about matters affecting their safety at work. They should make available to safety representatives information within the employer's knowledge which is necessary to enable safety representatives to fulfil their function. In particular this should include

* information about plans and performance of the undertaking and any changes proposed in so far as they affect the health and safety at work of their employees
* information of a technical nature about hazards in respect of machinery, plant, equipment, processes, systems of work and substances in use at work
* information relating to the occurrence of any accident or notifiable industrial disease
* results of any measures taken by the employer in the course of checking the effectiveness of his health and safety policy.

The code also includes guidance on the functions of safety committees. Among others these should
• give consideration to the circumstances of individual accidents, and the study of accident statistics and trends, so that reports can be made to management on unsafe and unhealthy conditions and practices, together with recommendations for corrective action
• examination of safety audit reports on a similar basis
• consideration of reports and factual information provided by inspectors
• consideration of reports which safety representatives may wish to submit
• assistance in the development of works safety rules and safe systems of work
• periodic inspection of the workplace, its plant, equipment, and amenities
• a watch on the effectiveness of the safety content of employee training, and on the adequacy of safety and health communication and publicity in the workplace
• provision of a link with the appropriate inspectorates of the enforcing authority.

Contract of employment
Finally there are aspects of occupational health and safety which can affect the contract of employment. The most important are
• Effect of breach of safety rules by employee—can he be dismissed?
• Effect of breach of safety rules by employer, i.e. can employee claim to have been 'constructively dismissed'?
• Is the employer under a duty to warn of potential risks of the employment? In relation to vibration white finger, it was held in *White* v *Holbrook Precision Castings* [1985] 1 RCR 215 that such a duty did exist.
• What is the position of an employer with regard to employee risks enhanced by pregnancy? In such a case, sex discrimination law may be invoked.
• The effect of sickness and absence from work of an employee and its effect on the contract of employment. Once again very many factors can affect the outcome in each individual case, reference should be made to the individual's contract, collective bargaining agreement and common sense. However, if the employee is away for long enough, then his contract of employment may be said to be *frustrated*, i.e. comes to an end by itself. The employee may be refused his job back and the employer will not be liable for a claim for unfair dismissal, as he is not in fact dismissing the employee. Again, legal advice should be sought.

11   Some legal background to occupational health

11.7 Industrial relations aspects of occupational health

In 1977 the Employment Appeal Tribunal in the case of *Egg Stores (Stamford Hill) Ltd.* v *Leibovici* laid down guidelines as to the matters to be considered when an employer tries to avoid a claim for unfair dismissal or redundancy pay by arguing that the contract had been frustrated:
- the length of the previous employment
- how long it had been expected that the employment would continue
- the nature of the job
- the nature, length and effect of the illness
- the need of the employer for the work to be done and the need for a replacement to do it
- the risk to the employer of acquiring obligations in respect of redundancy payments or compensation for unfair dismissal to the replacement employee
- whether wages have continued to be paid
- the acts and statements of the employer in relation to the employment, including the dismissal of, or failure to dismiss, the employee
- whether in all circumstances a reasonable employer could be expected to wait any longer.

In respect of these, and most other matters pertaining to civil liability, advice should be sought from a solicitor.

Suspension on medical grounds
The Employment Protection (Consolidation) Act 1978 sections 19, 20 and Sch. 1 provides that, where an employee is suspended from work on medical grounds in consequence of (i) any requirement imposed by or under the provision of any enactment, or (ii) any recommendation contained in a code of practice issued under the HSW Act, which is a provision specified in Schedule 1 of the Employment Protection (Consolidation) Act, then the employee is entitled to be paid during the suspension for a period of up to 26 weeks. The purpose of the payments is to compensate workers removed from their normal work, owing to the risks involved. An employee has to comply with any reasonable requirements specified by the employer. Currently the provisions under Schedule 1 are
- Control of Lead at Work Regulations 1980
- Ionising Radiations Regulations 1985, Reg. 16.

12 Education

12.1 Introduction, 321

12.2 Occupational medicine, 322

12.3 Occupational health nursing, 325

12.4 Occupational hygiene, 330

12.5 Centres of study for occupational health, hygiene and nursing, 331

12.1 Introduction

During the last decade, there have been major changes in the educational requirements for training in occupational health in Britain. Occupational medicine and occupational hygiene specialities have both moved away from *ad hoc* appointments of untrained practitioners and a disparate array of postgraduate courses. Both specialities have now developed independent bodies to oversee the quality of trainees. Though neither body actually organizes training courses, they do supervise the formal examinations and have set themselves up as the arbiters of approved training and specialist accreditation. Occupational health nursing education is validated by the National Boards for Nursing, Midwifery and Health Visiting.

National Council for Vocational Qualifications

Arising out of the White Paper 'Working Together—Education and Training' published in 1986 the Government of the day set up the National Council for Vocational Qualifications (NCVQ). This body, incorporated as an independent company limited by guarantee, sets out to define and recognize national standards of competence for most trades and professions practising within the United Kingdom. Their aims are defined in their publication *Criteria and Procedures* (1989).

Their method is to appoint a 'Lead Body' in the discipline which defines the levels of competence required for the various subject areas that make up the field. The Industry Lead Body consists of representatives from Government, industry and employees who are the 'users' of the products from the examination boards, training and teaching establishments. These organizations themselves do not sit on the Lead Body but are represented on advisory working groups. At present (1991) there are over 150 lead bodies.

Occupational health and safety is represented in two ways: firstly by ensuring that there is an input to all industry lead bodies and secondly by having its own lead body to define the competences of the practitioners in the field. The Health and Safety Executive is in the forefront of both of these approaches, in the first case by having nominated members of staff reponsible for an input to other industry lead bodies and in the second by providing the chairmanship and secretariat for the Occupational Health and Safety Lead Body (OHSLB). The composition of the OHSLB is as follows:

Confederation of British Industry (CBI)	4
Trades Union Congress (TUC)	2
Local Authorities (LA)	1

Health and Safety Executive (HSE)	1
Employment Department (Training,	
Enterprise and Education Directorate—(TEED)	1

The various disciplines that make up the field of occupational health, e.g. nursing, medicine, hygiene, are represented on one or more working groups.

For the future, there are moves afoot both in the UK and in the rest of Europe to 'harmonize' training requirements and 'standardize' vocational training courses, one aim of which is to allow free movement of appropriately qualified professionals across national boundaries.

12.2 Occupational medicine

To practise occupational medicine in the UK one requires only a basic registrable medical qualification. However, for many years, the more competent occupational health services have employed physicians with postgraduate qualifications, such as the Diploma in Industrial Health (DIH) or the Master of Science in Occupational Medicine.

Although in law nothing has changed, in practice much has altered. In 1978, the Faculty of Occupational Medicine (FOM) was established at the Royal College of Physicians of London. This development followed directly upon the recognition by the Royal College of the speciality of occupational medicine. This seemingly mundane event was, in fact, of momentous importance because, in the wake of the decision of the College, certain events were put automatically in train:

• the establishment of an academic focus
• the need to maintain standards of competence and ethics
• the need to formalize future speciality training
• the need to establish registers of accredited specialists.

The FOM is now responsible for academic standards, ethics and competence, whilst a Standing Advisory Committee for Occupational Medicine has been formed within the Joint Committee of Higher Medical Training (JCHMT). Between them, they oversee

• the establishment and maintenance of suitable trainee posts
• the accreditation of specialists
• the approval of training programmes
• the examination of physicians at the end of their training period.

At the outset, the FOM created three grades of membership

• associate member
• full member
• fellow.

12.2 Occupational medicine

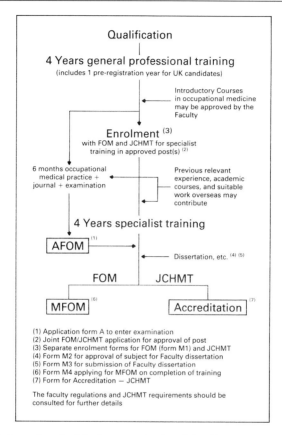

Fig. 12.1 An outline of training in occupational medicine. (Reproduced with kind permission from the FOM.)

Associate membership was designed as a qualification for doctors who are working part-time and/or working towards full membership, which is broadly equivalent to accreditation. Whilst the JCHMT insists that higher specialist training should only start after a period of general medical training, normally leading to the MRCP (UK) or the MRCGP, the FOM felt that, although such qualifications would be advantageous, it was not a sufficient test of competence. Thus, the examination for associate membership (AFOM) was specifically developed. The training routes for budding occupational physicians in the UK are represented diagrammatically in Fig. 12.1.

Fellowship would be by peer review and election, though it is possible to achieve membership and fellowship 'by distinction'. In February 1991, the figures for the Faculty of Occupational Medicine were

12.2 Occupational medicine

Honorary fellows	31
Fellows by distinction	20
Fellows	217
Members by distinction	49
Members	723
Associates	685
Total	1725

From 1981 various 'back door' clauses have been available to achieve associate and full membership by means of experience in the field but, with the first examination for associate membership in July 1982, the 'back door' routes slowly closed. From 1985, the only route is through examination for associate membership and then onwards, if desired, in the normal way to full membership.

The associate membership examination consists of
• a written examination (two papers)
• a clinical examination
• an oral examination
• a projected materials section consisting of slides with questions and multiple-choice answers
• a 'journal' of 2000 words concerning the candidate's activities during his/her training period.

During training, either formally or informally, the candidate should have covered the following academic areas
• principles of occupational health
• structure and function of occupational health services
• effects of work on health
• effects of health on work
• sources of information and use of computers
• clinical occupational medicine
• occupational hygiene
• occupational safety
• occupational toxicology
• epidemiology and statistics
• occupational health law
• occupational health, environmental health and public health.

Associate membership of the FOM will, in future, be considered to be the basic postgraduate qualification in occupational medicine and is likely to become the criterion for appointment in a part-time post of some import. Full-time practitioners will, however, often wish to reach consultant grade

12.3 Occupational health nursing

Table 12.1 Guidelines of the JCHMT Standing Advisory Committee on occupational health subjects for accreditation

Compulsory	Recommended	Acceptable
Surveillance of individuals or groups of workers	Occupational skin disease	Research
	Occupational lung disease	Toxicology
Management of workers developing work-related ill health	Work physiology	Pneumoconiosis
	Ergonomics	Aviation medicine
Assessment of disability and fitness for work	Epidemiology	
		Naval medicine
Assessment and advice on physical and psychological aspects of the working environment	Liaison with community and environmental health	Radiation
	Occupational health law	Health education
Involvement in all aspects of industrial organization		
Personal responsibility for the management of a department or sub-department of occupational health		

posts in the health service or in industry and this will require further training to full membership level and the establishment of their accredited specialist status. During the process of accreditation, a number of options are available to enable candidates to subspecialize and thereby encompass the discipline of aviation or naval medicine. For this purpose, the Standing Advisory Committee on Occupational Medicine of the JCHMT has laid out guidelines on the basis of those areas of the subject to be considered 'compulsory', 'recommended' and 'acceptable'. These are listed in Table 12.1.

Details of the Regulations for the Diplomas of Associateship and Membership of the FOM (revised May 1990) are available from the FOM.

12.3 Occupational health nursing
With the introduction of the statutory framework for nurses in 1979, an in-depth review was carried out into educational preparation for nurses, midwives and health visitors which was known as 'Project 2000'. The results of this review have brought about changes in educational strategies within the profession. It highlighted three major influencing factors affecting education:

12.3 Occupational health nursing

---

1   The need to reorientate nurses towards community care, to promote self-care and independence in a wide range of settings, e.g. home, school, work, as well as clinical hospital settings.
2   The need to prepare nurse practitioners to contribute to the planning, assessment and development of services, including their own future role.
3   The need to educate nurses, midwives and health visitors to cope with the rapid changes taking place within the sphere of health care, and to prepare them to adapt to meet the needs of the people they care for.

'Project 2000' proposed a single level of registered practitioner who should be competent to be an autonomous practitioner. The education would take place in both institutional and non-institutional settings and there should be a common foundation programme of two years' duration with the main focus being on health not illness, and for the learning to take place in various settings, not just hospitals.

Following the common foundation programme the individuals select one of four branch programmes—mental illness, mental handicap, nursing adults or children. This branch programme would last for a further year, at the end of which, on successful completion, the individual would emerge as a Registered Practitioner in that branch, e.g. RMN, RNMH, RGN, RSCN.

Through this programme nurse education has been able to seek academic accreditation together with professional approval. 'Project 2000' programmes have opened new avenues—learners at the successful completion of the programme will have achieved study at accepted, accredited or diploma level. Some programmes will be at degree level. Post-registration education for specialist practice will also require academic accreditation to the minimum of diploma level and that accreditation will need to encompass the concept of credit accumulation.

Credit Accumulation and Transfer Schemes (CATS) This is a system for providing credit for learning. Credit accumulation is a system where learning from a variety of experiences and subjects can be credited. Credit transfer means that credit acquired from another course can be transferred across to others. This should exclude repetitive learning.

All occupational health nursing courses in the future will be validated by the National Boards and CNNA/ Universities as a joint process. They will be validated in the future at diploma level and higher educational establishments are being asked to upgrade all their current programmes. The occupational health nursing diploma will be a recordable qualification by the UKCC, as is the present certificate.

In 1990 the occupational health nursing syllabus was developed using the framework of the 'Hanasaari' model for occupational health nursing

12.3 Occupational health nursing

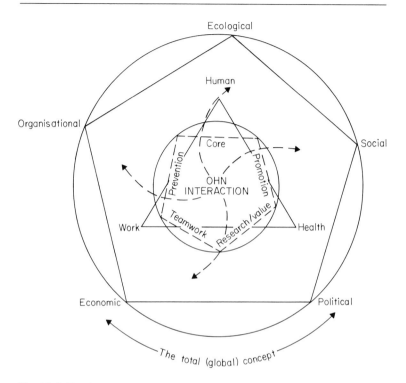

Fig. 12.2 The Hanasaari conceptual model for occupational health nursing (R. M. Alston *et al.*, 1989, Workshop for Occupational Health Nursing, Hanasaari, Finland).

(Fig. 12.2). The Hanasaari model was developed at a workshop for occupational health nurses held at Hanasaari, Finland. This was produced to allow for flexibility in occupational health nursing practice. The occupational health nursing concepts identified in the model can be developed in relation to professional practice and standards, education and management.

The model is presented in the following ways:

The total environment concept

The general environmental system which incorporates health and safety aspects is represented by the large outer circle—a global concept. Within the outer circle the influences which have global effects, which in turn may affect health, are represented by economic, political, social, ecological and organizational factors.

12.3 Occupational health nursing

The human, work and health concept
This is represented by the human, work and health triangle and operates within the total environment, aspects of the total environment having a significant effect on workplace health. Political and social policies for example will either expand or contract the development of occupational health. Organizational cultures and strategies may exert a stronger more direct influence on the human, work and health triangle.

Occupational health nursing interaction
Occupational health nursing is presented in the centre of the model. Interaction has been used to illustrate the areas identified by the groups as the role of the occupational health nurse.

Details of the National Boards for England, Wales and Northern Ireland occupational health nursing courses (with kind permission from the Occupational Health Nursing Syllabus, 1990, English National Board for Nursing, Midwifery and Health Visiting, London)

Requirement for becoming a qualified occupational health nurse
Those wishing to qualify as an occupational health nurse must:
1 be a nurse on Part I of the Register of the United Kingdom Central Council for Nursing, Midwifery and Health Visiting (UKCC) or have obtained temporary registration on Part I of the Register of the UKCC by virtue of an acceptable overseas qualification.
2 satisfactorily complete a course approved for the attainment of a qualification in occupational health nursing acceptable for recording on the Register of the UKCC.
3 produce evidence of satisfactory general education, preferably a minimum of 5 subjects in the General Certificate of Education at Ordinary level.
    Some educational institutions may require students to pass an entrance test in addition to or instead of the educational attainments listed; some may require additional educational qualifications.
    In addition, some institutions may stipulate post-registration experience requirements.

General information
This syllabus has been designed to allow for innovation in curriculum development. This is in line with the policy of the ENB and with current curricular issues in education. The trend is to move away from being prescriptive on course content. It will also enable educational establishments to take into account local needs and institutional expertise.

12.3 Occupational health nursing

Course aim
To facilitate understanding of occupational health nursing concepts and the application of specialized care in organization/community practice settings.

Learning outcomes
The learning outcomes which follow are required to prepare the nurse to apply knowledge and skills to meet the health needs of individuals and groups which arise from their interface with the work situation:

1   The understanding of ethical issues in the nursing profession and the responsibilities which these impose on occupational health nursing practice.

2   The identification of socio-economic, political and cultural factors which affect health at work.

3   The recognition of factors which contribute to, and those which adversely affect, the physical, mental and social well-being of those who work, and provide appropriate care.

4   An understanding of the requirements of legislation relevant to the practice of occupational health nursing.

5   The identification of the educational needs of clients in order to participate in health promotion, taking into account community strategies.

6   The use of interpersonal and communication skills appropriate to effective occupational health nursing practice: the use of channels of referral for matters not within the occupational health nurse's sphere of competence.

7   The identification of chemical, physical, biological, organizational, psycho-social and ergonomic hazards which affect health.

8   The ability to apply a process approach to the application of care in a work situation.

9   The ability to operate effectively within a team and participate in an inter-disciplinary approach to effect professional occupational health practice.

10   An understanding of epidemiology and research methods to support the development of occupational health nursing practice.

11   An understanding of global environmental issues and their effects on health.

12   An understanding of organizational theories and the ability to use appropriate management strategies in the application of care in the workplace.

Courses available
The occupational health nursing course is of not less than 320 hours' tuition by approved lecturers.

The course may be either (i) spread over a period of not more than one academic year of full-time study; or (ii) spread over a period of not more than two academic years of day release study of not less than 65 days.

Courses are based in Institutions of Higher or Advanced Further Education and include both theoretical tuition and practical experience.

12.4 Occupational hygiene

Qualifications in the United Kingdom are under the control of the British Examination Board in Occupational Hygiene (BEBOH), the examination board for the Institute of Occupational Hygienists.

This Board offers no courses but sets those examinations outlined below.

Preliminary certificates
These are in the following subject areas; examination is by a multiple-choice question paper plus essay.

1   Principles of British occupational hygiene practice.
2   General principles of workplace control.
3   Assessment of air pollution from industrial sources.
4   Toxic metals.
5   Microbiological dangers of occupations.
6   Noise and vibration.
7   Thermal environment.
8   Ionizing radiation.
9   Fires and explosions.
10   Dust causing pneumoconiosis and byssinosis.
11   Harmful dusts, liquids, vapours and mists.
12   Lightning and non-ionizing radiation.
13   Asbestos.
14   Principles of making an assessment under the COSHH Regulations 1988.

Certificate of Operational Competence
Entry requirements: 5 'O' levels, including English language and 2 'A' levels
    in science subjects or equivalent, acceptable, qualifications, plus 3
    years' experience in the comprehensive field of occupational hygiene.

Examination: two written papers plus an oral examination. Note that
candidates holding 6 appropriate prelimary certificates are exempt from
written papers.
Status: Registered Operational Hygienist, eligible for grade of Licentiate of
the Institute of Occupational Hygienists.

Diploma of Professional Competence
Entry requirements: honours degree in science or engineering or
equivalent, plus 5 years' approved experience in the field of
occupational hygiene.
Examination: four written papers plus an oral examination. Candidates who
have spent at least 10 years in full-time occupational practice may be
examined by published work, plus a set paper written in a specific time
and an oral examination.
Status: Registered Professional Hygienist, eligible for grade of Member of
the Institute of Occupational Hygienists.

12.5 Centres of study for occupational health, hygiene and nursing

Course centre	Department	Course title*	Duration
MEDICINE/HEALTH			
University of Aberdeen	Environmental and Occupational Medicine	MSc (Occ. Health)	Modular PT or FT
University of Birmingham	Institute of Occupational Health	M. Med. Sci. (Occ. Health)	1 yr FT
		AFOM course	2 yr PT
University of Dundee	Community and Occupational Medicine	Certificate Industrial Health	3 months FT
University of Edinburgh	Occupational Health	MSC	1 yr FT
University of Manchester	Occupational Health	AFOM course, MSc	2 yr PT or Distance Learning Course
University of Newcastle	Occupational Health and Hygiene	AFOM course	2 yr PT
University of Surrey	Robens Institute	MSc	1 yr FT
		Diploma	Modular PT

12.5   Centres of study for occupational health, hygiene and nursing

Course centre	Department	Course title*	Duration
HYGIENE			
Aston University	Occupational Health and Safety	Diploma, Safety and Health Certificate, Safety and Hygiene	6 months 1 yr FT
Polytechnic of the South Bank	Institute of Environmental Science and Technology	BSc, Occupational Hygiene	3 yr FT
University of Birmingham	Institute of Occupational Health	Short courses (various)	
University of Bradford	School of Environmental Science	BTech, Environmental Science	4 yr SW
University of Newcastle	Occupational Health and Hygiene	MSc, Occupational Hygiene	1 yr FT
University of Manchester	Medical School	MSc, Occupational Hygiene	2 yr PT
BEBOH			
Doncaster Metropolitan Institute of Higher Education	Science and Maths	P-Certificates in Occupational Hygiene	2 weeks each
Hinton and Higgs Training			
Institute of Occupational Medicine, Edinburgh			
National Occupational Hygiene Service, Manchester			
Polytechnic of the South Bank	Centre for Industrial Safety and Health		
Polytechnic of Wales	The Technology Centre		
Portsmouth Polytechnic	School of Architecture		
Salford University	Building and Environment		
University of Aberdeen	Environmental and Occupational Medicine	MSC Occupational Hygiene	Modular— PT or FT

12 Education

12.5 Centres of study for occupational health, hygiene and nursing

Course centre	Department	Course title*	Duration
University of Surrey	Robens Institute	Short courses (various)	
University of Ulster at Jordanstown			

Note: Other organizations also run P-Certificate courses from time to time. For the latest list contact: BEBOH, Suite 2 Georgian House, Great Northern Road, Derby, DE1 1LT, tel: 0332 298087.

NURSING

Centres for study for OHNA, OPN and OHNC exist throughout the UK and are subject to approval from the National Boards.

England	
Bristol	University of Bristol
	Wills Memorial Building
	Queens Road
	Bristol BS8 1HR
Guildford	University of Surrey
	Guildford
	Surrey GU2 5XH
Ipswich	Suffolk College of Higher and Further Education
	Rope Walk
	Ipswich
	Suffolk IP4 1LT
London	Institute of Advanced Nursing Education
	Royal College of Nursing
	20 Cavendish Square
	London W1M 0B
Luton	Luton College of Higher Education
	Park Square
	Luton
	Bedfordshire LU1 3JU
Manchester	Manchester Polytechnic
	Elizabeth Gaskill Campus
	Hathersage Road
	Manchester
Portsmouth	Highbury College of Technology
	Cosham
	Portsmouth PO6 2SA
Teesside	Teesside Polytechnic
	Middlesbrough
	Cleveland TS1 3BA

12.5 Centres of study for occupational health, hygiene and nursing

Twickenham	West London Institute of Higher Education
	Gordon House
	St Margarets Road
	Twickenham
	Middx TW1 1PT
Wolverhampton	The Polytechnic
	Wulfruna Street
	Wolverhampton WV1 1LY
N. Ireland	None available at present
Scotland	Robert Gordon's Institute of Technology
	School of Health and Social Work
	352 King Street
	Aberdeen AB9 2TQ
Wales	None available at present

FT, full-time; PT, part-time; SW, sandwich course (one year in industry).
*Many of these establishments run short courses from one day to one month in duration.

13 Sources of information

13.1 Introduction, 337

13.2 Further reading, 337

13.3 Useful addresses, 340

13.1 Introduction

The field of occupational health and safety is so wide that no single individual can expect to be knowlegeable in every aspect. In the area of toxicological effects of chemical substances, for example, nobody could be expected to know details of the thousands of chemicals in common use. Therefore access to a good library in occupational health containing general and specialist texts is essential and, where possible, the facility to consult electronic data bases is useful.

A valuable source of information is the Health and Safety Executive, who issue a wide range of printed material from free leaflets and information sheets to priced guidance notes normally available through HMSO.

There is a large range of textbooks available from the large generalist texts to ones specializing on the many individual topics that make up this multidisciplinary subject.

Electronic data bases are extremely valuable as they are able to contain vast amounts of material, far in excess of what is reasonable to purchase in printed texts, and they have the advantage of being updated several times per year. For this purpose, it is necessary to invest in an IBM-compatible desk top computer plus a CD-ROM reader, as the data sources are supplied on compact discs of the same size as music discs. Each compact disc can, apparently, hold the equivalent of 88 000 sheets of A4 printed material.

It is impossible to cover every textbook and data source in this chapter but the authors offer the following sources from their experience as being useful for further reading.

13.2 Further reading

General texts in occupational medicine/disease

Adams R.M. (1990) *Occupational Skin Disease*, 2nd ed., W.B. Saunders, Philadelphia.

Berensen A. (Ed.) (1980) *Control of Communicable Diseases in Man,* 12th ed., American Public Health Association, Washington DC.

Department of Health and Social Security (1981) *Industrial Diseases: A Review of the Schedule and the Question of Individual Proof,* Cmnd. 8393, HSMO, London.

Edwards F.C., McCallum R.I., Taylor P.J. (Eds) (1988) *Fitness for Work—The Medical Aspects,* Oxford Medical Publications, Oxford.

Employment Medical Advisory Service (1977) *Occupational Health Services: The Way Ahead,* HMSO, London.

Essex-Cater A.J. (1979) *A Manual of Public Health and Community Medicine,* 2nd ed., John Wright, Bristol.

13.2 Further reading

Harrington J.M. (Ed.) *Recent Advances in Occupational Health* **2** (1984) and **3** (1987), Churchill Livingstone, Edinburgh.

Howard J.K., Tyrer F. (1987) *Textbook of Occupational Medicine,* Churchill Livingstone, Edinburgh.

Hunter's Diseases of Occupation, Raffle P.A.B., Adams P., Baxter P.S., Lee W.R. (Eds) (1992) 8th ed. Hodder & Sloughton, London.

International Labour Office (1988) *Encyclopaedia of Occupational Health and Safety,* 3rd ed., ILO, Geneva.

Levy B.S., Wegman D.H. (Eds) (1988) *Occupational Health,* 2nd ed., Little Brown & Co., Boston.

MacDonald J.C. (Ed.) (1980) *Recent Advances in Occupational Health* **1**, Churchill Livingstone, Edinburgh.

National Institute of Occupational Safety and Health (1977) *Occupational Diseases: A Guide to Their Recognition,* NIOSH, Cincinnati.

Parkes W.R. (1991) *Occupational Lung Disorders,* 3rd ed., Butterworth & Co. London.

Perkin H. (1969) *The Origins of Modern English Society* (1780–1880), Routledge and Kegan Paul, London.

Rom W.N. (1988) *Environmental and Occupational Medicine,* 2nd ed., Little Brown & Co., Boston.

Rosenstock L., Cullen M.R. (1986) *Clinical Occupational Medicine.* W.B. Saunders Co., London.

Shaw E.R. (Ed.) (1981) *Prevention of Occupational Cancer,* CRC Press, Chicago.

Tyrer F.H., and Lee K. (1985) *A Synopsis of Occupational Medicine,* 2nd ed., John Wright, Bristol.

Waldron H.A. (Ed.) (1989) *Occupational Health Practice,* 3rd ed., Butterworth & Co., London.

Waldron H.A. (1990) *Lecture Notes on Occupational Medicine,* 4th ed., Blackwell Scientific Publications, Oxford.

Ward-Gardner A. (Ed) (1979) *Current Approaches to Occupational Medicine,* John Wright, Bristol.

Zenz C. (1980) *Developments in Occupational Medicine,* Yearbook Publications, Chicago.

Zenz C. (1988) *Occupational Medicine,* 2nd ed., Yearbook Publications, Chicago.

Occupational safety law

Fife I., Machin E.A. (1990) *Redgrave's Health and Safety,* Butterworth & Co., Sevenoaks.

Ridley J. (Ed.) (1990) *Safety at Work,* 3rd ed., Butterworth–Heinemann, London.

Toxicology

Amdur M.O., Doull J., Klaassen C.D. (Eds.) (1991) *Casarett and Doull's Toxicology,* 4th ed., Pergamon Press, Oxford.

Dillon H.K., Ho M.H. (Eds.) (1987) *Biological Monitoring of Exposure to Chemicals—Organic Compounds,* John Wiley & Sons, Chichester.

Finkel A.J. (1983) *Hamilton and Hardy's Industrial Toxicology,* 4th ed., John Wright, Bristol.

Health and Safety Executive. *Toxicity Reviews—*Series of critical reviews of the

13 Sources of information

13.2 Further reading

toxicology of substances considered by the Health and Safety Commission's acts.

Muir G.D. (Ed.) (1986) *Hazard in the Chemical Laboratory,* 4th ed., Royal Society of Chemistry, London.

National Institute of Occupational Safety and Health and the Occupational Safety and Health Administration (1990) *Occupational Health Guidelines for Chemical Hazards,* NIOSH, Cincinnati.

Patty's Industrial Hygiene and Toxicology (1978—1987) Clayton G.D., Clayton F.E., Vols. 1, 2A, 2B, 2C, 3A and 3B, John Wiley & Sons, Chichester.

Proctor N.H., Hughes J.P. (1988) *Chemical Hazards of the Workplace,* 2nd ed., Lippincott Co., Philadelphia.

Sax N.I. (1989) *Dangerous Properties of Industrial Materials,* Vols. 1, 2 and 3, 7th ed., Van Nostrand Reinhold, New York.

Epidemiology

Checkoway H., Pearce N.E., Crawford-Brown D.J. (1989) *Research Methods in Occupational Epidemiology,* Oxford University Press, Oxford.

Hennekens C.H., Buring J.E. (1987) *Epidemiology in Medicine,* Little Brown & Co., Boston.

MacMahon B., Pugh T.F. (1970) *Epidemiology: Principle and Methods,* 2nd ed., Little Brown, Boston.

Monson R. (1990) *Occupational Epidemiology,* 2nd ed., CRC Press, Florida.

Occupational hygiene

American Conference of Government Industrial Hygienists (1986) *Documentation of TLVs and BEIs,* 5th ed., ACGIH, Cincinnati.

American Conference of Government Industrial Hygienists. *Threshold Limit Values for Chemical Substances and Physical Agents and Biological Exposure Indices for 1990–1991,* ACGIH, Cincinnati.

Ashton, I., Gill F.S., (1991) *Monitoring for Health Hazards at Work,* 2nd ed., Blackwell Scientific Publications, Oxford.

BOHS Technical Guide No. 7 (1987) *Controlling Airborne Contaminants in the Workplace,* Science Reviews Ltd, Leeds.

Croner's Handbook of Occupational Hygiene. Looseleaf publication, regularly updated. Croner Publishing, Kingston upon Thames.

Health and Safety Executive (1991) *Guidance Notes EH 40/91, Occupational Exposure Limits 1991,* HMSO, London.

Ho M.H., Dillon H.K. (Eds) (1991) *Biological Monitoring of Exposure to Chemicals—Metals,* John Wiley & Sons, Chichester.

National Institute of Occupational Safety and Health (1984) *Manual of Analytical Methods,* 2 vols., NIOSH, Cincinnati.

Waldron H.A., Harrington J.M. (Eds) (1980) *Occupational Hygiene,* Blackwell Scientific Publications, Oxford.

Occupational health nursing

Harris C. (1984) *Occupational Health Nursing Practice,* John Wright, London.

Radford J.H. (Ed.) (1990) *Recent Advances in Nursing. Occupational Health Nursing,* Churchill Livingstone, Edinburgh.

13.3 Useful addresses

Professional organizations

British Occupational Hygiene Society, Secretary: Dr J.G. Underwood Suite 2,
  Georgian House, Great Northern Road, Derby DE1 1LT.
Chartered Institution of Building Services (incorporating what were formerly the
  Institution of Heating and Ventilation Engineers and the Illuminating Engineering
  Society), Delta House, 222 Balham High Road, London SW12 9BS.
Ergonomics Society, Hon. Secretary: K.E. Coombes, Dept. of Human Sciences,
  University of Technology. Loughborough LE11 3TU.
Faculty of Occupational Medicine, Royal College of Physicians, 6 St Andrew's Place,
  London NW1 4LE.
Institute of Occupational Hygienists, Dept. Secretary: T.G.E. Gillanders, Suite 2,
  Georgian House, Great Northern Road, Derby DE1 1LT.
Institute of Occupational Safety and Health, Secretary: J. Barrel, 222 Uppingham
  Road, Leicester LE5 0QG.
Institute of Radiation Protection, 64 Dalkeith Road, Harpenden, Herts, AL5 5PW.
Royal College of Nursing Society of Occupational Health Nursing, 20 Cavendish
  Square, London W1M 0AB.
Society of Occupational Medicine, 6 St Andrew's Place, London NW1 4LE.
United Kingdom Central Council (UKCC) for Nursing, Midwifery and Health Visiting,
  23 Portland Place, London W1N 3AF.
  English National Board for Nursing, Midwifery and Health Visiting, Victory House,
  170 Tottenham Court Road, London W1P 0HA.
  National Board for Nursing, Midwifery and Health Visiting for Scotland, 22 Queen
  Street, Edinburgh EH2 1JX.
  Welsh National Board for Nursing, Midwifery and Health Visiting, 14th Floor, Pearl
  Assurance House, Greyfriars Road, Cardiff CF1 3AG.
  National Board for Nursing, Midwifery and Health Visiting for Northern Ireland, RAC
  House, 790 Chichester Street, Belfast BT1 4JE.

Government organizations

Building Services Research and Information Association, Old Bracknell Lane West,
  Bracknell, Berks RG12 4AH.
Department of the Environment. (a) Building Research Establishment: Building
  Research Station, Garston, Watford WD2 7JR. (b) Fire Research Station:
  Borehamwood, Hertfordshire WD6 2BL. (c) Warren Spring Laboratory: Gunnels
  Wood Road, Stevenage, Herts SG1 2BX.
Health and Safety Executive. (a) Directorate of Information and Advisory Services: St
  Hugh's House, Stanley Precinct, Bootle, Merseyside L20 3QY. (b) The
  Occupational Medicine and Hygiene Laboratories: 403–405 Edgware Road,
  Cricklewood, London NW2 6LN. (c) Health Policy Division and Medical Services
  HQ, Baynards House, 1 Chepstow Place, Westbourne Grove, London W2 4TF.
Laboratory of the Government Chemist (Department of Trade and Industry), Queens
  Road, Teddington , Middx TW11 0LY.
Medical Research Council. (a) Headquarters: 20 Park Crescent, London W1N 4AL.
  (b) Environmental and Epidemiology Unit: South Academic Block, Southampton

13.3 Useful addresses

General Hospital, Southampton SO9 4XY. (c) Toxicology Unit: Woodmasterne Road, Carshalton, Surrey SM5 4EF.

Trade and research associations

Agricultural and Food Research Council, Wiltshire Court, Farnsby St., Swindon SN1 5AT.

British Ceramics Research Association, Queens Road, Penkhull, Stoke on Trent, Staffordshire ST4 7LQ.

British Rubber Manufacturers Association Ltd., 90–91 Tottenham Court Road, London W1P 0BR. Birmingham Branch: Health Research Unit, Scala House, Holloway Circus Queensway, Birmingham B1 1AQ.

Chemical Industries Association, Kings Buildings, Smith Square, London SW1P 3JJ.

Welding Institute, Abington Hall, Cambridge CB1 6AL.

Advisory and information services

Aston University, Health and Safety Unit, Dept. of Mechanical and Production Engineering, Aston Triangle, Birmingham B4 7ET.

Bath University, European Documentation Centre, The Library, Claverton Down, Bath, Avon BA2 7AY.

British Rail Research and Development Department, Railway Technical Centre, London Road, Derby DE2 8UP.

Central Toxicology Research Laboratory, ICI Ltd., Alderley Park, Macclesfield, Cheshire SK10 4TJ.

Industrial Health Ltd., The Old Surgery, Queen Victoria Road, Newcastle upon Tyne NE1 4HL.

Institute of Aviation Medicine, Farnborough, Hants. GU14 6SZ.

Institute of Naval Medicine, Alvestoke, Gosport, Hants, PO12 2DL.

Institute of Occupational Health, University of Birmingham, Edgbaston, Birmingham B15 2TT.

Institute of Occupational Medicine, 8 Roxburgh Place, Edinburgh EH8 9SU.

Institute of Sound and Vibration Research, University of Southampton, Southampton SO9 5NH.

National Occupational Hygiene Service Ltd., Skelton House, Manchester Science Park, Lloyd Street North, Manchester M15 4SH.

O'HEAL, 23 Highcroft Industrial Estate, Enterprise Road, Horndean, PO8 0EW.

Poison Centre, New Cross Hospital, Avonley Road, London SE14 5ER.

Regional Toxiocology Laboratory, Dudley Road Hospital, Dudley Road, Birmingham B18 7QH.

Robens Institute of Industrial and Environmental Health and Safety, University of Surrey, Guildford, Surrey GU2 5XH.

South Bank Polytechnic, Institute of Environmental Engineering, Borough Road, London SE1 0AA.

Thompson Laboratories, The Stocks, Cosgrove, Milton Keynes MK19 7JD.

United Kingdom Atomic Energy Authority, Harwell Laboratory, Harwell, Didcot, Oxon, OX11 0RA.

University of Bradford, Radiation Protection and Occupational Hygiene Service, Bradford, West Yorks, BD7 1DP.

University of Manchester, Centre for Occupational Health, Medical School, Stopford Building, Oxford Road, Manchester M13 9PT.

13 Sources of information

13.3 Useful addresses

University of Newcastle, Division of Environmental and Occupational Medicine, Medical School, School of Health Care Sciences, Framlington Place, Newcastle upon Tyne NE2 4HH.
University of Strathclyde, Department of Civil Engineering and Environmental Health, Coalville Building, 48 North Portland St., Glasgow G1 1XN.
Wolfson Institute of Occupational Health, University of Dundee, Level 5, Medical School, Ninewells, Dundee DD1 9SY.

International sources

American Conference of Government Industrial Hygienist, P.O. Box 1937, Cincinnati, Ohio 45211, USA.
Commission of the European Communities, 200 Rue de la Loi, B-1049 Brussels.
International Agency for Research on Cancer, 150 Cours Albert-Thomas, 69372 Lyon, Cedex 08, France.
International Commission on Occupational Health, Sec./Treasurer: Professor J. Jeyaratnam, Dept. of Community, Occupational and Family Medicine, National University Hospital, Lower Kent Ridge Road, Singapore.
International Commission of Radiological Protection, c/o Dr H. Smith, Clifton Avenue, Sutton, Surrey SM2 5PU.
International Labour Office, CH-1211 Geneva 22, Switzerland.
International Occupational Safety and Health Information Centre (CIS), International Labour Office, 154 Rue de Lausanne, CH-1211 Geneva 22, Switzerland.
National Institute for Occupational Safety and Health, Robert A. Taft Laboratories, 4676 Columbia Parkway, Cincinnati, Ohio 45226, USA.
Occupatioanl Safety and Health Administration, US Dept. of Labor, Washington DC 20210, USA.
World Health Organization, CH-1211 Geneva 27, Switzerland.

Index

Access to Medical Records Act
(1988) 315
Accidents 219
Acquired immune deficiency syndrome
(AIDS) 222
Acrylamide 117, 139
Acrylonitrile 139–40
Acts of Parliament as source of law 297
Afterdamp 159
AIDS (acquired immune deficiency
syndrome) 222
Airborne contaminants
control of
displacement of pollutants by airflow
over worker 273
enclosing or shielding working
273–4
exposed population 241–2
fugitive (unpredictable)
emissions 242, 243
hardware form of 239, 240
periodic and continuous
emissions 242–4
software form of 239, 240
source of emission 240–4
transmission 241
see also Fans in ventilation systems;
Personal protection of worker;
Ventilation systems
occupational hygiene standards
for 59–62
actions when no standards exist
61–2
mixed exposures 60–1
standard-setting authorities 59–60
occupational hygiene survey
techniques 57–8
Alveolitis, extrinsic allergic 49, 99, 114
types 100 (Table 4.4)
American Conference of Government
Industrial Hygienists (ACGIH),
workplace environmental
standards produced by 59, 126
Aniline 140–1
Ankylostomiasis (hookworm) 114, 209
Anthrax 113, 209–10
Antimony (Sb) 127
Arsenic (As) 115, 127–8
Arsine (AsH_3) 127, 165–6

Asbestos 97, 99, 119, 121, 170–2, 228
Asbestosis 98 (Fig. 4.5), 99, 121,
171–2
Asthma 95–6, 120
Audiometry 198, 293
Australia, occupational health services
in 24

Barytosis 99
Benzene 36, 56, 116, 138, 141
Beryllium (Be) 117, 128–9
Biological agents of occupational
disorders 209–15
Biological Exposure Indices (BEI) 54, 56
Biological monitoring 53–5, 56
of effect 54–5, 55
of exposure 54–5
Blackdamp 159
Breach of statutory duty as cause for civil
action 313
Breathing apparatus 286–7
Brucellosis 114, 210
Byssinosis 96–7, 119, 174

Cadmium (Cd) 56, 117, 129
Canada, occupational health services
in 23
Cancer, occupational 226
carcinogenesis, chemical, theories
of 227–8
carcinogens, chemical
classification 228 (Table 8.1), 230
(Table 8.2)
epigenetic 227, 228 (Table 8.1)
genotoxic 227, 228 (Table 8.1)
carcinogens, Group I, list of
(International Agency for Research
on Cancer) 230 (Table 8.2)
carcinogens, occupational
characteristics of 228
known or suspected 228–32
historical perspective 226–7
lung cancer 97, 99, 117, 121, 122,
171
occupations presenting a carcinogenic
risk 230–1 (Table 8.3)
Carbon dioxide 159

Carbon disulphide (bisulphide) 36, 116, 142
Carbon monoxide 56, 159–60
Carbon tetrachloride 36, 142–3
Carboxyhaemoglobin 56, 160, 162
Carcinogens *see under* Cancer, occupational
Cardiovascular system 103–5
Case law 299–300
Chemical agents 125
  inorganic chemicals *see individually named inorganic chemicals*
  monitoring of workplace environment 126
  organic chemicals 138 *see also individually named organic chemicals*
  toxic gases 158, 165–6
    chemical asphyxiants 159–62
    irritants 162–5
    simple asphyxiants 158–9
  workplace environmental standards 125–6
Chest radiograph 93–4, 98 (Fig. 4.5)
Chlamydiosis 115
Chlorinated naphthylenes 106, 116, 143
Chlorine (Cl$_2$) 163
Chloroform 119, 143–4
Chokedamp 158
Chromium (Cr) 57, 130
Civil liability for occupational injuries and illnesses
  breach of statutory duty 313
  negligence 312–13
  statutes of limitation 314
Coal dust 97–9, 169–70
Codes of practice 17, 298, 306
  Control of Substances Hazardous to Health Regulations, compliance with 34, 239–40
  International Labour Organisation as issuing body 304
  safety representatives and safety committees 316–17
Consumer Protection Act (1987) 42, 301, 313
Contract of employment affected by occupational health and safety factors 317–18

suspension on medical grounds 318
Control of Substances Hazardous to Health Regulations (1988) (COSHH) 29, 32–6, 125, 239, 274, 298
  assessment of health risks 33, 36, 41–5
  pro-forma for 37–41
  data sheets 42–3, 44–5
Cotton dust 96, 173–4
Credit accumulation and transfer schemes (CATS) in learning 326

Dampers in extract systems 255–6
'Danish painters syndrome' 102
Data Protection Act (1984) 315
Data Sheets 42–3, 44–5
DDT (1,1,1-trichlorobis(chlorophenyl) ethane) 144
Decompression sickness 111, 202–3
Dermatitis 105–7, 120
  allergic contact 106–7
  primary irritant contact 105–6
Dinitrobenzene 145
Dinitrophenol 145–6
Disciplines and professions related to occupational health 4–11
Displacement of airborne pollutions by airflow over worker 273
Divers 47, 202–3
Drivers
  medical examinations 47, 50, 78
  whole body vibration 204
Dusts 61, 96–9, 115, 119, 174–5, 242–3
  asthma 95–6, 120
  chest radiograph 93
  deposits in ducts of ventilation systems 250–1, 255–6
  dermatitis 120
  extrinsic allergic alveolitis 99–100
  see also Asbestos; Coal dust; Cotton dust; Silica dust

Ear-muffs 289, 290, 291–3
Ear-plugs 289, 290

Education and training in occupational
health 231
centres of study 331–3
credit accumulation and transfer
schemes (CATS) 326
National Council for Vocational
Qualifications 321–2
occupational health nursing 7–8,
325–30
courses of National Boards 328–30
study centres 332–3
occupational hygiene 330–1
study centres 331–2
occupational medicine 322–5
study centres 331
Employment medical advisers 32, 34,
311–12
Employment Medical Advisory Service
(EMAS) 311–12
Employment Protection (Consolidation)
Act (1978) 318
Enclosing or shielding worker from
hazardous environment 273–4
Enforcement of health and safety
legislation in UK
see Health and safety legislation in UK,
enforcement of
Environment health in UK 16
Environmental Hygiene Guidance Notes
(EH) 58, 59, 60, 61, 126
Epidemiology 8–9
definition of 8, 62
health survey design 62–6
appraising a published study 65
shortcomings of epidemiological
method 65
Ergonomics 5, 9
Ethers 147–9
Ethics in occupational health
services 21, 315
Europe, occupational health in 22–3
European Community (EC)
directives 301, 302–3
'framework' directive (1989) 19, 22,
302–3
general duty of care 302–3
importance of 301
personal protective equipment
275

European Community (EC) law 297,
300–3
sources 301
Evaluation of workplace hazards see
Workplace hazards, evaluation of
Expatriates' health 77, 214–15
Extrinsic allergic alveolitis 49, 99, 114
types 100 (Table 4.4)
Eye and face protection 275–8

Factories Act 304, 313
Faculty of Occupational Medicine
(FOM) 4, 322–5
Fans in ventilation systems 253–6
axial flow fans 262–3
bifurcated 263
centrifugal fans 263–5
backward-curved 265–6
forward-curved 264–5
radial-bladed (paddle-bladed) 265
fan characteristic curves 261
fan pressures 260
matching of fan and system 266
power and efficiency definitions
260–1
propeller fans 261–2
Farmer's lung 99, 100, 114
Fletcher equation 249
Fluorine (F_2) 163–4
Food handlers, medical
examinations 47, 50, 51–2, 78
Food Safety Act (1990) 52
Formaldehyde 146, 220
Fume 95, 115–20, 175
from welding 137
Fume cupboard 245, 246, 250

Genitourinary system 102–3
Glanders 114, 210–11
Glutaraldehyde 220–1
Gonioma kamassi poisoning 116
Group health services 20
Guidance notes (by Health and Safety
Commission and Health and
Safety Executive) 299
Environmental Hygiene Guidance
Notes (EH) 58, 59, 60, 61, 126

Hanasaari model for occupational health
   nursing 326–8
Hand-arm vibration syndrome
   (HAVS) 204, 205
Hazard data sheets 42–3, 44–5
Health and Safety at Work Act (1974) 42,
   83, 297, 304–10, 313, 316
Health and Safety Commission 298,
   305–6
Health and Safety (Enforcement
   Authority) Regulations
   (1989) 306–10
Health and Safety Executive (HSE) 49,
   59, 126, 321
  enforcement of health and safety
    legislation 306, 307, 309–10
Health and safety legislation in UK,
   enforcement of 306–10
  enforcement notices 310
   appeals against 310
  enforcing authorities, jurisdiction of
   of Health and Safety Executive 306,
    307, 309–10
   of local authorities 306, 308–9
Health of health-care workers 219–23
  occupational hazards, examples
   of 220–3
Health surveillance 34, 36, 53, 55
  see also Medical examinations; Medical
   questionnaires
Hearing protection 289–93
Heat
  environmental monitoring 176
   air velocity 176
   clothing worn 181
   duration of exposure 182
   effective and corrected effective
    temperature 184–6, 186
   heat stress index 187
   kata thermometer 176, 178, 179
   metabolic rate (for different
    activities) 179, 182
   psychrometric chart 176, 177, 178,
    179
   radiant heat exchange 176, 179,
    180–3 (Fig. 6.5)
   wet bulb globe temperature 183
   wind chill index 187
  hazards of health-care workers 220

  health effects 175–6
Hepatitis A (infective hepatitis) 211
Hepatitis B (serum hepatitis) 211
  health-care workers' hazards 222
Hepatitis C 211
Hepatitis D (delta agent hepatitis) 211
Hippuric acid 56, 154
Hookworm (ankylostomiasis) 114, 209
Human immunodeficiency virus (HIV) 222
Hydatidosis 115
Hydrogen cyanide (HCN) 160
Hydrogen sulphide (H$_2$S) 161

Industrial Injuries Advisory Council
   (IIAC) 86
Industrial relations 312, 316–18
Industrial tribunals 300, 310
Industry Lead Body 321
International Agency for Research on
   Cancer (IARC) 229
International Labour Organisation
   (ILO) 18–19, 22, 303–4
International law 303–4
International occupational health 22–4
Ionizing radiation 199–201
  health-care workers' hazards 220
Isocyanates 49, 147

Japan, occupational health services in 24
Joint Committee on Higher Medical
   Training (JCHMT) 4, 322, 325
Judicial precedent as source of law
   299–300

Kata thermometer 176, 178, 179
Ketones 147–9

Laser radiation 201–2
  health-care workers' hazards 220
Law relative to occupational health
   10–11
  civil liability 312–14
   breach of statutory duty 313
   negligence 312–13
   statutes of limitation 314

Index

contract of employment affected by occupational health and safety factors 317–18

European Community law 297, 300–3
sources 301

industrial relations 312, 316–18

international law 303–4

medical ethics and confidentiality 21, 315

sources of, in UK 297–300, 306

LD_{50} (measure of toxicity) 90

Lead (Pb) 56, 91, 115, 130–2

Lead bodies, and vocational qualifications 321–2

Legge, Sir Thomas (1863–1932) 84

Legionnaire's disease 213–14

Leptospirosis 114, 211

Limitation Act (1980) 314

Liver, and occupationally related diseases 107–8, 109 (Table 4.6), 118

Local authorities
as enforcing authorities 306, 308–9
environment health responsibilities 16

Low back pain in health-care workers 222–3

Lungs
acute inflammation 94–5
cancer 97, 99, 117, 121, 122, 171
potential occupational allergens 96
see also Respiratory system

Malaria 212

Man-made mineral fibres (MMMF) 172–3

Manganese (Mn) 115, 132

Maximum exposure limit (MEL) 32, 33, 44, 59–60, 126

Medical audit 223–4

Medical ethics and confidentiality 21, 315

Medical examinations 34, 36
Employment Medical Advisory Service (EMAS) 311–12
medical records 21, 34, 55, 57
occupational history taking 48–9, 73
periodic 45–6, 53, 55, 71–2, 78–9
drivers (heavy goods vehicles; public service vehicles) 50, 78

executives 52, 53
food handlers 50, 51–2, 78
medical surveillance levels 53
women 51
young and old employees 50–1
post-sickness-absence examination 49–50
pre-placement 47–8, 66–71

Medical questionnaires (for completion by employees or applicants) 52, 74–7

Medical records 21, 34, 55, 57
Access to Medical Records Act (1988) 315

Medical surveillance levels 53

MEL (maximum exposure limit) 32, 33, 44, 59–60, 126

Meningococcal meningitis 222

Mental health of workers 223, 224–6

Mercury (Hg) 56, 115, 132–4, 221

Mesothelioma 98 (Fig. 4.5), 99, 119, 171

Methane 138, 158–9

Methods for the Determination of Hazardous Substances (MDHS) 126

Methyl alcohol (methanol) 149

Methyl bromide (bromomethane) 149–50

Methylene chloride (dichloromethane) 150

Microwave radiation 202

Miner's nystagmus 112

Mortality, occupationally related 87–8

National Council for Vocational Qualifications 321–2

National Health Service (NHS) 15–16, 18, 20
health of health-care workers 219–23

Negligence 299, 312–13

Nervous system, occupationally related disorders 100–2
central nervous system 101–2
peripheral nerves 100–1

Nickel (Ni) 134

Nickel carbonyl ($Ni(CO)_4$) 116, 134, 161–2

Nitric oxide (NO) 164
Nitrogen 158, 159
  oxides of 116, 164
Nitrogen dioxide (NO_2) 164
Nitrous oxide (N_2O) 164
Noise 187–8
  addition of sounds 189
  auditory health effects 197–8
  decibel, definition of 188
  decibel weightings 191
  health-care workers' hazards 220
  $L_{EP.d}$ in noise dose 192–4
  L_{eq} unit in noise dose 191–2
  noise dose 191–4
  noise ratings 190–1
  non-auditory health effects 198
  pressure in sound 8
  sound spectrum 189
  see also Noise at Work Regulations
    (1989)
Noise at Work Regulations (1989) 192,
    194–6, 289, 298
Nominal protection factor (npf) in
    respiratory protection 283, 288
Nurses
  role in occupational health 6–8
  see also Occupational health nursing
Nurses, Midwives and Health Visitors
    Act (1979) 7

Occupational diseases
  historical perspective 83–4
  mortality, occupationally related 87–8
  notifiable diseases 85–6
    see also Reporting of Injuries,
      Diseases and Dangerous
      Occurrences Regulations (1985)
  prescribed diseases 85–7
    listed schedule of 111–22
      (Appendix 4.1)
  target organs 91
    cardiovascular system 103–5
    genitourinary system 102–3, 104
      (Table 4.5)
    liver 107–8, 109 (Table 4.6)
    nervous system 100–2
    reproductive system 108–11

  respiratory system see Respiratory
    system
  skin 105–7
  toxicology, general principles 88–91
  workers with endocrinologically active
    pharmaceuticals 110–11
Occupational exposure limits (OEL) 126
  see also Maximum exposure limit
    (MEL); Occupational exposure
    standard (OES); Short-term
    exposure limit (STEL); Time-
    weighted average (TWA)
Occupational exposure standard (OES)
    32, 33, 44, 59, 60, 126
Occupational health, definition and
    general features 3–4, 11
Occupational Health and Safety Lead
    Body (OHSLB) 321–2
Occupational health ethics 21, 315
Occupational health nursing
  education, training, and qualifications
    in 7–8, 325–30
    courses of National Boards 328–30
    study centres 332–3
  general features 6–8
Occupational health services 15, 25
  aims and functions 18–19
  international 22–4
  UK 17–22
    ethics in 21, 315
    future of 21–2
    types of 19–20
Occupational hygiene
  airborne contaminants, standards
    for 59–62
  definition and general features 5–6
  education, training, and qualifications
    in 330–1
  study centres 331–2
  survey techniques 57–8
Occupational medicine
  education, training, and qualifications
    in 322–5
  study centres 331
  general features 4
Occupational Safety and Health
    Administration (OSHA),
    Washington, airborne
    contaminants, standards for 59

Orders in Council 300
Orf 115
Organophosphates 150–1

Personal protection of worker 274–5
 European legislation 275
 eye and face protection 275–8
 hearing protection 289–93
 respiratory protection 282–8
 skin and body protection 278–81
Phenol 36, 56, 151–2
Phosgene (carbonyl chloride)
  ($COCl_2$) 138, 142, 165
Phosphine (PH_3) 165–6
Phosphorus (P) 134–5
Platinum (Pt) 135
Pneumoconiosis 97–9, 119
 ILO/UC classification 93–4
Pontiac fever 214
Pregnancy, abnormal, occupational
  factors associated with 109–10
Pressure 202–3
 losses in ductwork systems 251–6
 of sound 188
Private medicine in UK 17
'Project 2000' in nursing 325–6
Psittacosis 213
Psychrometric charts 176, 177, 178,
  179

Q fever 115, 212

Radiation 198–202
 ionizing 199–201
  health-care workers' hazards 220
 non-ionizing 201–2
  lasers 202, 220
  microwaves 202
Regulations, as source of occupational
  health law 297–8, 306
Rehabilitation and resettlement 234–5
Repetitive strain injury (RSI) 232–4
Reporting of Injuries, Diseases and
  Dangerous Occurrences
  Regulations (1985) 49, 85, 86,
  125

Reproductive system, and occupationally
  related disease 108–11
Respirators 283–6
Respiratory system
 chest radiograph 93–4, 98 (Fig. 4.5)
 occupational lung disorders 94–100,
  117, 121, 122, 171
 personal protection equipment 282–8
 structure and function 91–3
Reynolds' number 252–3

Safety committees 316–17
Safety engineering 10
Safety representatives 316–17
Short-term exposure limit (STEL) 59, 126
Siderosis 99
Silica dust 97, 170
Silicosis 97, 98 (Fig. 4.5)
Single European Act (1988) 301–2
Skin, and occupationally related
  diseases 105–7, 120
 personal protection of worker (skin and
  body) 278–81
Sources of information on occupational
  health 337
 further reading 337–9
 useful addresses 340–2
Stannosis 99
Statutes of limitation 314
Statutory Instruments 298
STEL (short-term exposure limit) 59, 126
Stibine (SbH_3) 127, 165–6
Streptococcus suis infection 114
Stress 224–6
 in ambulance workers 223
 in junior hospital doctors working long
  hours 223
Styrene 152–3
Sulphur idoxide (SO_2) 162–3

Tetrachloroethane 116, 153
Tetrachloroethylene
  (perchlorethylene) 153–4
Thallium (TI) 136
Third world, occupational health services
  in 24
Threshold limit values (TLVs) 59, 126

Time-weighted average (TWA), in workplace environmental standards 59, 126
Toluene 56, 154
Toxic gases 158, 165–6
 chemical asphyxiants 159–62
 irritants 162–5
 simple asphyxiants 158–9
Toxicology, general principles 88–91
Treaty of Rome 297, 301–2
1,1,1-Trichloroethane 45, 56, 105, 155
Trichloroethylene 36, 155–6
Trinitrotoluene 156
Tuberculosis (TB) 114, 213
 health of health-care workers 221
TWA (time-weighted average) 59, 126

United Kingdom Central Council (UKCC) for Nursing, Midwifery and Health Visiting 7, 328
USA, occupational health in 23, 59

Vanadium (V) 136
Ventilation systems
 air cleaning 267
 dilution ventilation 256, 258–60
  aims (summary of) 259–60
 discharges to the atmosphere 267–8
 energy and cost implications 268–70
 extract ventilation 244
  aims (summary of) 259–60
  aspect ratio 245
  capture distance 245
  capture velocity 244–5
  dampers 255–6
  ducts and fittings 250–1
  face velocity 245
  fan see Fans in ventilation system
  low-volume, high-velocity extract systems 248
  pressure losses 251–6
  Reynolds' number 252–3
  static pressure 251–3
   suction inlets 245–8, 249–50
   system resistance characteristic 256
   transport velocity in ducts 251
   velocity pressure 251–3
 legal requirements on ventilation of workplaces 260
Vibration 203–5
 hand-arm 204, 205
 Whole body 204
Vinyl chloride 35, 36, 57, 156–7
Vitiligo 118

Walk-through surveys 29–32
Welding fumes 137
Workplace hazards, evaluation of 29
 airborne contaminants, occupational hygiene standards for 59–62
 biological monitoring 53–5, 56
  of effect 54–5, 55
  of exposure 54–5
 Control of Substances Hazardous to Health Regulations (1988) 29, 32–6, 298
  assessment of health risks 33, 36, 41–5
  data sheets 42–3, 44–5
  pro-forma for use in assessments 37–41
 health survey design, epidemiological approach 62–6
 occupational hygiene survey techniques 57–8
 walk-through surveys 29–32
 see also Medical examinations; Medical questionnaires
World Health Organisation 24

Xylene 157, 220

Zinc (Zn) 137
Zinc protoporphyrin 55, 56, 131